Advanced Structured Materials

Volume 63

Series editors

Andreas Öchsner, Southport Queensland, Australia
Lucas F.M. da Silva, Porto, Portugal
Holm Altenbach, Magdeburg, Germany

More information about this series at http://www.springer.com/series/8611

J.M.P.Q. Delgado · A. Gilson Barbosa de Lima
Editors

Drying and Energy Technologies

Springer

Editors
J.M.P.Q. Delgado
Department of Civil Engineering
University of Porto
Porto
Portugal

A. Gilson Barbosa de Lima
Department of Mechanical Engineering
Federal University of Campina Grande
Campina Grande, Paraíba
Brazil

ISSN 1869-8433 ISSN 1869-8441 (electronic)
Advanced Structured Materials
ISBN 978-3-319-37273-0 ISBN 978-3-319-19767-8 (eBook)
DOI 10.1007/978-3-319-19767-8

Springer Cham Heidelberg New York Dordrecht London

Softcover reprint of the hardcover 1st edition 2016

Printed on acid-free paper

Springer International Publishing AG Switzerland is part of Springer Science+Business Media
(www.springer.com)

Contents

Drying of Bioproducts: Quality and Energy Aspects

A.G. Barbosa de Lima, J.V. da Silva, E.M.A. Pereira, I.B. dos Santos and W.M.P. Barbosa de Lima

Abstract This chapter briefly focuses on the drying of wet bioproducts with particular reference to fruits, vegetables and grains. Different related topics in terms of drying foundations, dryer selection, product quality, energy savings, energy sources, energy efficiency, energy recovery, operating safety, environmental impact, and advanced drying techniques are presented and discussed. The study confirm drying as a highly energy-consuming process, one of the major source of pollutant emissions, and one dehydration technique that strongly affect product quality under different aspects such as color, flavor, appearance, aroma, losses of nutrients and vitamins, and many others physical, chemical, structural, and nutritional quality parameters.

Keywords Drying · Dryer · Fruits · Vegetables · Quality · Energy · Efficiency · Heat recovery

A.G. Barbosa de Lima (✉) · J.V. da Silva · W.M.P.B. de Lima
Department of Mechanical Engineering, Federal University of Campina Grande, Av. Aprígio Veloso, 882, Bodocongó, Campina Grande, PB 58429-900, Brazil
e-mail: antonio.gilson@dem.ufcg.edu.br

J.V. da Silva
e-mail: jvieira7@gmail.com

W.M.P.B. de Lima
e-mail: wan_magno@hotmail.com

E.M.A. Pereira
Federal Institute of Education, Science and Technology of Paraiba, Picuí, PB, Brazil
e-mail: evaldomarcos@bol.com.br

I.B. dos Santos
Department of Physics, State University of Paraiba, Campina Grande, PB, Brazil
e-mail: ivoneetebs@gmail.com

J.M.P.Q. Delgado and A.G. Barbosa de Lima (eds.),
Drying and Energy Technologies, Advanced Structured Materials 63,
DOI 10.1007/978-3-319-19767-8_1

1 Drying: Why and How?

On the basis of the states of water (liquid and/or vapor) in a material is possible to classify it as unbound (free) and bound water. Bound water corresponds to the water present as a liquid solution within the solid or even trapped in the microstructure of the solid. Unbound water in a porous material is the water in excess. In general, unbound water in a material can be removed by using mechanical and thermal methods. In this chapter, the mechanical removal of water will not be considered.

Drying is perhaps the oldest and most common unit operation (thermal method) found in several industrial that occur in the chemical, pharmaceutical, paper, mineral, polymer, ceramic, textile, food and agricultural industries. It is also one of the most complex and least understood process which involves simultaneous and coupled heat, mass, and momentum transport in a matter media. The aim is to reduce the moisture content of the product for different purposes. However, most of the deteriorative processes that take place in foodstuffs are influenced by the temperature, and concentration and mobility of water inside the food. Thus, adequately to control the process plays an important role.

Many perishable food products of great nutritional importance for human diet and nutrition, such as fruits and vegetables (rich in vitamins and minerals), are dehydrated in the food industry. The loss of nutrients and functional attributes varies with regard to the type, condition and time of dehydration process and the sensitivity of specific food components. During dehydration the product is generally above room temperature but well below sterilization temperature. The added heat and exposure time of the product at elevated temperature affect the nutrient quality of the food products, especially that very sensitive to heating. Further, perishable food materials are more stable at low than at high moisture content, and low storage temperature [1]. Thus, the goal of the food preservation is to convert perishable foods such as grains, fruits and vegetable in stabilized products, consequently reducing postharvest losses, in order to extend storage self life and to quality enhancement.

Drying can be realized using equipment so called dryer. There are different types of dryers such as band, flash, conveyor, drum, fluid-bed, vacuum, rotary, spray and cyclone dryers. They are of diverse sizes, shapes and varieties. Because of particular characteristics of the material to be dried (for examples, moisture content, particle-size distribution, bulk density, hardness, dust content, flow characteristics, color, odor, appearance, consistency, dissolution or rewetting behavior, caking tendency, etc.), each group of material is dried in a specific dryer (or class of dryers), in ways batch or as continuous operation [2].

Selection of a particular dryer is one the most complex tasks. It is also one of the most least understood operations because of the difficulties and deficiencies in descriptions of the phenomena of simultaneous—and often coupled and multiphase—transport of heat, mass, and momentum in wet solid media including dimensions variations and others physical and chemical phenomena associated to drying process. Generally, it has been made by specialists based on scientific knowledge, past experience, and frequently own intuition. The following data must be assessed for selection of a suitable dryer [2–9]:

(a) Method of operation (batch or continuous).
(b) Method of heat supply (convective, contact, thermal radiation).
(c) Suitable energy sources.
(d) Direction of heat and wet material flow.
(e) Mode and pattern of drying medium flow (Free or forced flow).
(f) Throughput and evaporative capacity (production capacity).
(g) Physical, chemical, and biochemical properties of the wet feed.
(h) Desired product specifications.
(i) Upstream and downstream processing operations.
(j) Moisture content of the feed and product.
(k) Maximum allowable product temperature.
(l) Drying kinetics; moist solid sorption isotherms.
(m) Quality parameters (physical, chemical, biochemical).
(n) Safety aspects (fire hazard and explosion hazards, toxicological properties).
(o) Value of the product.
(p) Need for automatic control.
(q) Corrosion aspects.
(r) Pressure in dryer.
(s) Flow pattern, transport method and mixing of solids.
(t) Cost of dryer.

Despite of the importance of drying, the following questions must be made: Drying provokes some irreversible effect in the product? How to do it efficiently in terms of time, energy cost, and investment and product quality? The answer of these questions will be given in the topics following.

2 Quality Aspects in Drying

For obtain good quality of a biological product implies that the dried product undergoes several physical, chemical or biological changes to yield a product of desired specifications.

In many industries, there are different process step, among them we can cite separation and concentration of the components mixtures. Generally, fruit and vegetables in nature have high moisture content ranging from 60 to 90 % (wet basis). Overall post-harvest losses of fruit and vegetables in developing countries are estimated at about 20–50 % of the production which justify its dehydration until a specified and safe moisture content though drying process, for example, under viewpoint of storage [10]. Further, in the food processing industry water is added to perform several of the operations. After these operations water must be removed through evaporation and drying processes [11]. In the ceramic industry, during of manufacturing process of building materials, water is added to clay, in order to confer plasticity of the mixture and thus, facilitate the molding process of pieces (clay product such as brick, roof files, etc.). After the forming step and before the firing in kilns, water contained in wet product (moisture content 10–40 % on dry basis) must be removed by drying process, using chamber-dryers or tunnel dryers until moisture content ranging from 0.5 to 6 % on dry basis.

Drying is one of the oldest and most important preservation methods for reduction of moisture content of foods or other heat sensitive, biologically active products. Drying incurs material to several changes, which sometime are irreversible and responsible for low quality of dry materials. The extent of quality damaging depends on the time-temperature-water content history during drying [12]. Then, the goal is the maximization of product quality through process control.

With respect to the drying process, quality deterioration mechanisms may be classified as nutritional, physical or chemical [13, 14]. Quality of dried bioproducts, such as foods, grains, pharmaceutical preparations and other materials depends on compositions of the material, conditions of drying and conditions of processing and storage. The following most important mechanisms causing quality loss are: shrinkage, collapse, surface changes, crystallization of salts, sugars, and polymer, causing stickiness and caking, maillard reaction, vitamin oxidation, enzymatic and non-enzymatic browning, lipid oxidation, color loss, microbial variations, and protein denaturation. These phenomena are related to mobility of components of the material being dehydrated [15]. All changes depend on water activity, temperature and exposure time [16, 17].

Many bioproducts, such as foodstuffs and wood, darken when heated or exposed to sunlight. Controlling this change in appearance is an important issue in maintaining the quality of dried products [18]. According to Keey [18], drying of green sapwood causes a significant enrichment of the melanin precursors (low-molecular sugars and nitrogeneous compounds).

Heating and moisture removal of the fruits and vegetables, for example, result in many changes in properties of the tissue subjected to drying. Plant tissue softness when heated due to loss of turgor and complex changes of the cell wall matrix polysaccharides. Loss of water and concentration of soluble add rigidity to the cell wall and the outer layer of drying material acquire considerable mechanical strength [19], and thus changes in porosity is verified.

Mechanical resistance of the plant material, which consist of cells and fissures, restricts the extent of volume changes occurring during dehydration (increases rigidity) due to moisture removal, and thus, observed shrinkage is usually smaller than that predicted by theory (shrinkage is linearly related to water content, and equal to the volume of evaporated water) [20].

According to Suzuki et al. [21] and Yamamoto [22], many protein solutions are heat sensitive and easily denaturated at high temperatures; thus, one of the useful methods to keep protein activity and to protect it of the drying induced stresses and thermal inactivation during storage is freeze-dry protein solutions containing sugars. However, when adequate drying conditions and suitable stabilizers additives are used, it is possible to dry protein solutions with reduced thermal denaturation through convective drying [22].

For drying of some biomaterials, such as lipase, their inactivation appeared to be correlated with changes of the moisture content, while the inactivation of alkaline phosphatase and catalase is affected by drying agent temperature only [23, 24].

Sugar recrystallization in foods after drying is another problem. The sugars after drying are generally in the amorphous glassy state as they are dried until low moisture. These amorphous sugars are hygroscopic. Thus, if they are exposed to environment with high relative humidity, caking and crystallization will occur resulting in state changes forming either a sticky texture or it develops a hard texture [25].

Freeze drying (lyophilization) is a low temperature drying process and it has been used to convert solutions of labile products into solids of sufficient stability (drug manufacturing process) for distribution and storage, and thus, it is less likely to cause degradation than high temperature process, such as spray drying. However, while many proteins suffer little degradation during freeze drying, some proteins, such as multimeric proteins, often undergo significant conformational change and subsequent degradation during the freeze drying process [26].

Intermittent drying has been used widely for agricultural material and others heat-sensitive materials due to its benefit is terms of energy saving, time saving and higher quality of the final product after drying. For example, in grains, intermittent drying reduces thermo-hydrical stresses and fissures, thus increasing product quality after drying and reducing drying time. Intermittency strategies can include drying heat source, air mass flowrate, air velocity, air pressure and air temperature variations, where these parameters are periodically turned on and turned off, or surfer periodic variations, or they are controlled by external parameters [27, 28].

The diffusion of volatile aroma compounds is very important in food processing and storage, and an understanding of the mechanism of mass transfer will help in maintaining the food quality. On the basis of high relative volatility, high losses of volatiles would be expected during evaporation and drying of food products. However, those components may be retained at a relatively high percentage in the dried product because of the presence of soluble and insoluble solids in the food. Retention of aroma compounds depends on the concentration and nature of the food solids. During the fast evaporation in liquids food or wet solids foods, thermodynamic equilibrium is not reached with the very volatile components [29].

In general, the following quality changes during drying and storage of fruits can be cited [30]:

(a) Loss of vitamins (vitamins A and C);
(b) Loss of natural pigments (carotenoids and chlorophylis);
(c) Discoloration due to enzymatic or nonenzymatic browning;
(d) Oxidative degradation and flavor loss;
(e) Irreversible damage to the texture (shrinkage, slow cooking, and incomplete rehydration);
(f) Loss of water reducing microbiological spoilage.

Fruits and vegetables play an important role in human diet and nutrition as sources of vitamins and minerals.

In this chapter is stated as the concept of quality and energy of dried or rehydrated products are not simple. Further, these concepts are different for the industry and consumer. For consumers quality of dried foods must obey various criteria related mainly to appearance (color, gloss, shape and aspect), taste (taste itself, aroma, texture), convenience of use (convenience of instruction), rapidity of rehydration and of dissolution, ease of package opening, constancy of apparent density, keeping qualities, possibility of use (whole or pieces, complete or in part of the quantity, no stickiness), composition (nature and concentration of additives, content in calories and in fats, absence of residues of fertilizers or pesticides), microbiology (absence of pathogenic microorganisms, limited counts of others) [31].

For the industry the following factors of dried food quality can be cited: quality of raw material (tast aroma, appearance, water content, composition, absence of fertilizers or pesticide residues, microbiology, regularity, etc.), pretreatments (washing, peeling, cutting, sorting, blanching, incorporation of additives, such as SO_2, NaCl, etc., osmotic treatment, liquid concentration), drying conditions (product temperature transient history, thermal gradient and homogeneity of the temperature in the particle and bed of particle, presence/absence of particle stuck to hot surfaces, method of mechanical handling of the product, presence of oxygen, sorting after drying), packaging and storage conditions (type and quality of packaging material, water activity of product relative humidity of atmosphere to storage and packaging, cooling before packaging, duration and temperature of storage) [31].

In the optimization of a drying process, total cost and product quality are competing parameters. According to Bruin and Luyben [32] the following objectives in the drying of food materials can be given:

(a) Increased stability.
(b) Decreased transport and storage costs via weight reduction and/or volume reduction.
(c) Process economy: minimal product loss; higher heat and mass transfer rates, use of cheap energy source, use of cheap regeneration separating agent, minimal solids handling, easy operation, simple drying equipment, low environment impact.
(d) Good product quality: provided product structure and color, controllability of density, rapid and easy rehydration or redispersion, minimal chemical and biochemical degradation, selective solvent removal, simple packing and storage requirements, product without contamination.

3 Energy Aspects in Drying

With energy consumption ranged from 8 to 12 % of the industrial energy use, drying is considered as the one of the most energy intensive unit operations [33]. Thus, in many industrial drying processes, a lot of energy is wasted [34].

Thus, to reduce energy consumption per unit of product moisture, it is necessary to examine different methodologies to improve the energy efficiency of the drying equipment [35].

For obtain a product fully dried often requires a great deal of energy. Thus, sometimes is sufficient to dry a product to specific moisture content. This procedure reduces energy costs and increases both process efficiency and productivity of the dried products.

When a wet porous solid is subjected to thermal drying, three processes occur simultaneously: (a) Transfer of energy (mostly as heat) from the surrounding to evaporate the water at the solid surface; (b) Transfer of internal liquid water to the surface of the solid and its subsequent evaporation; (c) Transfer of energy to interior of the solid which is responsable to heating and evaporating of water.

Energy transfer as heat from the surrounding environment to the wet porous solid can occur as a result of convection, conduction, or radiation and in most cases as a result of a combination of these effects.

It has become apparent in recent years that energy resources, especially natural gas and oil, are limited. Consequently, all industrial sectors in all parts of the world need to identify more efficient methods of energy utilization. Despite periodic fluctuations, there is a tendency for energy costs to increase and, consequently, in many cases energy should become an important element of innovations to drying practice and changes in equipment technology.

The potential for energy reduction can be considered from several points of view, the most important being [36]:

(a) Methodological: Development of general methods for reducing energy consumed.
(b) Technological: Utilization of secondary energy sources, reduction of heat losses to the atmosphere, and the like.
(c) Socioeconomic: Stimulation of search for potential energy sources.
(d) Organizational: Coordination of processes aimed at an economical use of energy.
(e) Ecological: Focusing attention on solutions safe for the natural environment.

Thus, to reduce energy consumption per unit of product moisture, it is necessary to examine different methodologies to improve the energy efficiency of the drying equipment.

The drying behavior of an individual particle, or of a collection of particles in a dryer, depends on the geometrical and physical characteristics of the solid. We notice that, in these processes the particles will vary in shape and properties, as a consequence also vary in drying behavior (moisture migration, heating, shrinkage and dilation rates).

In present day, because of the high cost of the dryers, they consume a significant part of the total energy used in manufacturing processes, 12 % on the average and the its great application in many industrial processes and scientific interest, to study the drying process plays an important role. Then, development of the new technologies for the drying sector, improvement of the product and rationalization of energy consumption is crucial.

4 Energy Efficiency of the Drying and Dryer

Dryers are one of the major consumers of energy on a global scale and thus contributing significantly to the production of greenhouse gases (GHG). GHG emissions are generally calculated based on their CO_2 equivalent emissions [37]. According to Baker [38], a typical convective dryer (85 % of all industrial dryers) consumes around five times its capital cost in energy during its life time. Then, to develop an understanding of how much energy is being consumed (expended usefully or wasted) is crucial. For this, a heat and mass balance of the wet product and air provide a useful means for determine efficiency of the dryer.

Drying is an energy intensive operation that easily accounts for up to 15 % of all industrial energy usage, often with relatively low thermal efficiency in the range of 25–50 %.

In term of energy saving, we can define two major ratios to compare the combined overall drying operation with the convective one [39]:

(a) Time saving

$$\text{Time saving} = \frac{\text{Convective drying duration} - \text{Combined drying duration}}{\text{Convective drying duration}} \tag{1}$$

(b) Energy saving

$$\text{Energy saving} = \frac{\text{Convective drying energy consumption} - \text{Combined drying energy consumption}}{\text{Convective drying energy consumption}} \tag{2}$$

How to calculate energy efficiency? In general thermal energy efficiency can be determined by using heat and mass conservation equations as applied to air and wet material, or obtained as a product of partial efficiencies which represent external and internal factors that affect energy consumption in a dryer. These two manners for energy efficiency calculation can be used to obtain instantaneous or overall energy efficiency.

The literature has reported that heat supplied to the dryer is used for moisture removal (free water evaporation, capillary-bond water removal), heating of the wet material and dryer structure, compensation of heat losses for surroundings, local super heating of the water vapor, and overheating of the already dried layers of material otherwise necessary to maintain the required temperature gradient and thus sufficient heat transfer rates [40].

There is a wide variety of drying materials and types of equipment, so it is not possible to propose a generalized method for a standardized determination of dryer efficiency. The energy performance of a dryer and of a drying process is characterized by various indices such as volumetric evaporation rate, surface heat losses, steam consumption, unit heat consumption, and energy (thermal) efficiency.

There are different manner for calculation the performance of the dryer. Following, some of them are given. The specific energy consumption can be calculated as follows:

$$\epsilon = \frac{\text{Amount of water evaporated}}{\text{Energy used}} \tag{3}$$

The water evaporated is calculated from the drying curve and the energy input is the sum of the energy requirements. In the convective drying calculation of the energy requirements of the entire drying system, allowance must be done for the energy necessary to heat the drying air until the drying temperature and any heat recovery measures used [41].

Drying has been used in all types of agricultural products, including cereals, fruits, vegetables, and spices. It consumes around 0.38–0.63 GJ/ton of grain [37, 42].

According to reported literature, the average values of specific energy consumption obtained for the dryers in the ceramics, chemicals and food sectors were 3.98, 5.41, and 4.88 GJ/ton water evaporated, respectively [43].

The energy efficiency can be taken as:

$$\eta = \frac{\text{Energy required}}{\text{Energy supplied}} \tag{4}$$

Thus, the energy efficiency can be determined as overall, instantaneous, or it may be only for the drying chamber, excluding other peripheral energy requirements.

In processes involving electromagnetic waves, such as microwave and infrared, radiation and total efficiencies can be defined as follows [44]:

$$\eta_{\text{rad}} = \frac{\text{Total power radiated from the emitter}}{\text{Eletric power to the emitter module}} \tag{5}$$

and

$$\eta_{\text{Total}} = \frac{\text{Power obsorbed by the wet material}}{\text{Eletric power to the emitter mod ule}} \tag{6}$$

The total energy supplied to the wet product is related to latent heat of evaporated water, heat of sorption, sensible heat and convective losses from the material. For example, in the infrared, sun, solar and microwave drying processes the energy requirements is associated to the power absorbed by the material.

The use of solar energy in drying can significantly reduce the use of fossil fuels, and thus, reducing pollutant emissions. The daily efficiency of a solar collector is defined as the ratio between the energy supplied by the solar collector and the total incident radiation in a whole day of sun exposure, as follows [45]:

$$\bar{\eta} = \frac{\int_t q dt}{A \int_t I dt} \tag{7}$$

Raghavan et al. [37] defined the energy efficiency for convective dryers as follow:

$$\eta = \frac{T_{in} - T_{out}}{T_{in} - T_{env}} \tag{8}$$

where T_{in} and T_{out} correspond to the inlet and outlet drying medium temperature, and T_{env} represents the ambient air temperature. Thus, the maximum value of T_{out} is the wet bulb temperature. Equation (8) is associated with the drying chamber only. So, any other energy inputs and losses are not considered.

On the basis of the explanation above the following instantaneous energy efficiency can be given [40, 46] as follows:

$$\epsilon = \frac{\text{Energy used for evaporation at time t}}{\text{Input energy at time t}} \tag{9}$$

For a time interval, the cumulative energy efficiency can be determined as follows:

$$\bar{\epsilon} = \frac{1}{\Delta t}\int_0^t \epsilon(t)\, dt \tag{10}$$

Instantaneous and cumulative efficiencies which incorporate other sink such of energy as to breaking of the moisture-material bonds and moisture transportation to material surface is proposed by Menshutina et al. [40]:

$$\epsilon' = \frac{\text{Energy used for evaporation at time t}}{(\text{Input energy} - \text{Output energy with ontlet gas}) \text{ at time t}} \tag{11}$$

and

$$\bar{\epsilon}' = \frac{1}{\Delta t}\int_0^t \epsilon'(t)\, dt \tag{12}$$

A most complete definition for energy efficiency must include heat supplied to the dryer for moisture removal (free water evaporation, capillary-bond water removal), heating of the wet material and dryer structure, and heat losses for surroundings by convection and radiation, power of the fan (to flows the air) and power to product transportation. Thus, the following equation for calculation of the energy efficiency is proposed:

$$\eta = \frac{\dot{m}_{a\,in}(y_{a\,out} - y_{a\,in})h_{lv}}{\dot{m}_{a\,in}(y_{a\,out} - y_{a\,in})h_{lv} + \dot{m}_{a\,out}h_{a\,out} + \dot{Q}_{loss} + \dot{m}_p c_p T_{p\,out} + \dot{W}_{fan} + \dot{W}_p} \quad (13)$$

where $\dot{m}$ is the mass flow rate, y represent the absolute humidity, h is the enthalpy, c is the specific heat, $\dot{W}$ represent the power, and h_{lv} is the latent heat of vaporization. The subscripts in, out, p and a represent inlet, outlet, product and air, respectively.

Because the growing interest in energy savings and minimization of pollutant emissions (greenhouse gas) it is crucial to analyze different drying technologies, in order to save energy, provide high quality products post processing according to environment production laws [40].

We can notice that all those criteria cited before were proposed based on certain theoretical foundations; they have played important roles in the performance analysis and comparison of drying system and can be continually used in certain circumstances. However, from the view point of drying technology development, those criteria cannot be labeled as complete or rigorous, since these criteria only put view on the energy utilization efficiency without considering the drying quality requirement and environment impact, which is obviously against the general tendency of the drying technology progress [47].

According to Kemp [48], energy consume minimization in drying equipment may be achieved by one of the following methods:

(a) Reducing the inherent energy requirement for drying, e.g. by dewatering the feed.
(b) Increasing the dryer energy performance, by reducing heat losses, total airflow rate or batch times.
(c) Using the outlet energy between the dryer and surrounding through heat exchange.
(d) Heat recovery within the dryer system, between hot and cold streams.
(e) Using low–grade, lower–cost heat sources to supply the heat requirement.
(f) Combined heat and power; co-generate power while supplying the heat requirement to the dryer.
(g) Use of heat pumps to recover waste heat to provide dryer heating.

As an example, drying at lower temperature can be more efficient than that at higher temperature or yet combination of two or more drying techniques can often offer greater energy savings than applications using each drying mode separately. Hybrid or combined drying technologies include implementation of different modes of heat transfer, two or more stages of the some or different type of dryer [49].

5 Heat Recovery and Advanced Drying Techniques

Energy aspects are strongly related with the sustainable development. Thus, the use of energy is also closely connected with global environment problems, particularly climate change. In this sense, the increased use of biofuel is an effort to reduce the greenhouse gas emissions (CO_2 represents 80 % of the greenhouse gas emissions) and to increase capacity of the dryers [50].

Drying is a highly energy-consuming process and one of the major sources of atmospheric emissions in industrial operations. For heating of the drying air, conventional energy sources, such as fossil fuel (natural gas, propane, diesel and kerosene), electric energy; coal; waste materials, and renewable energy (solar and biogas), have been used for many years [36, 51].

As the fossil fuel becomes less available and more expensive, the ratio between drying costs/overall production costs will increase continuously. Thus, innovations in drying with actions related to energy efficient drying systems, air pollution control, and quality product preservation post-drying plays an important role [52–54].

Improvement in energy efficiency of the conventional drying equipment can include [2, 4, 5, 37, 55]:

(a) The shortening of drying time and dryer length.
(b) Supplying of heat precisely to interior of the dryer.
(c) Good dryer thermal isolation (less heat losses).
(d) Reduced drying agent consumption (low air flowrate).
(e) Energy recovery of outlet gas and/or hot dried product.

The main heat (sensible and/or latent) recovery methods are:

(a) Use of indirect gas/gas or gas/liquid heat exchange.
(b) Recycle of part of the exhaust air to the inlet air.
(c) Scrubbing of exhaust gas with cold water for obtain warm water.
(d) Using a heat pump to convert energy at a low level from exhaust air in energy at a high level and return it to the drying system.
(e) Combination of different methods.

On the basis of the arguments cited above, diverse new drying technologies have been cited in the literature, which include [55–57]:

(a) Use of superheat steam in direct dryer.
(b) Use of indirect heating (conduction).
(c) Use of combined (or integrated) heat transfer modes (dielectric or radiation with convection and/or conduction).
(d) Use of two—stage or multistage dryers.
(e) Use of intermittent heat transfer.
(f) Use of volumetric heating (microwave, radiofrequency, infrared, induction).
(g) Use of novel combustion technologies (pulse combustion for flash drying).
(h) Use of novel gas/solid contacts (two—dimensional, intermittent or rotating spouted beds, spout—fluidization, impinging streams).

(i) Design of flexible, multiprocessing dryers.
(j) Combination of different dryer types (pneumatic–fluid bed, spray–fluid bed, spray–conveyor dryer).
(k) Use of nonconventional drying (supercritical drying from native solvent, ultrasonic drying).

How related in previous sections, for choice of an appropriated drying energy system or comparing between different dryers depends on the drying energy consumption and equipment costs, drying time and mainly product quality involved in them. The total cost of drying must include the operating cost (energy consumption) and fixed (capital or fabrication) cost (costs of fan and motor, and the drying system structure) [51, 58].

Energy consumption basically is related to heating the air, and to transport the air and product through the dryer. Thus, heating, air flow and product feed control is crucial for energy saving in drying.

Conventional dryers used in the drying of bioproducts usually have utilized hot air under forced convection. Preheating of the drying air increases the heat transfer to the wet product and consequently increasing product temperature, and drying rate, which decreases the total drying time.

Industrial dryers often operate at a low thermal efficiency due to the drying air exits the drier at a high temperature and at a very low relative humidity. Thus, a small fraction of the exhaust air energy may eventually be recycled, using a single heat exchanger on the exhaust stream, to preheat the drying air [59–61]. Then, we need to develop new ecologically-friendly thermal energy utilization systems [62]. In this sense, heat pump dryer appear as an alternative method for energy-efficient drying. In this drying equipment energy of the warm and moist air from the dryer outlet, instead of venting off, it is moved to the condenser by the evaporator and compressor. The drying air recovery the heat from the condenser and uses it to dry the wet porous material into the dryer [63, 64].

In general, dryer exhaust gases are loaded with both, vapor and dust. Thus, condensation and formation of fouling layers can occur on cold heat-transfer surfaces reducing, in turn, hardly heat exchangers efficiency used for energy recovery. Then, knowledge of fouling mechanisms under the specific conditions of dryer exhaust gases is crucial for optimization of a design and efficient operation of heat recovery system [65].

6 Concluding Remarks

Drying of wet porous solids can be defined as a simultaneous and coupled heat, mass and momentum transfer process, often including dimensions variations, and is considered a technique with low energy efficiency. Information related in this chapter proves that to obtain good quality product with minimized energy consumption, an appropriated selection of drying equipment and operating parameters is crucial. The appropriated selection of a dryer depends on the installation and operation costs, product quality, operating safety, and convenience for installations, and many others characteristics. Further, thermal drying provokes significant physical, chemical, and nutritional changes in bioproducts, specially grains, fruits, and vegetables (thermo-sensitive materials). Damage intensity is determined in terms of drying parameters and composition, shape, size, and nature of the bioproduct.

For drying optimization different point of view such as methodological, technological, socioeconomic, organizational, and ecological have been carefully commented. Thus, we state that the knowledge of the drying behavior, quality of dehydrated products and energy efficiency is very important to determine the total drying time, and to confirm the use of a specified drying process in terms of energy saving and costs.

Results reported in this research indicated that controlling drying-air parameters (temperature, airflow rate and relative humidity) during the process has great potential for improving both efficiency and performance of the dryer, and product quality post-drying.

Under the aspect of energy efficiency, because of the wide variety of wet porous materials and types of drying equipment, herein different ways to calculate energy efficiency in drying and diverse new drying techniques are reported. The study shows that actual (conventional) drying system has used different energy sources, including fossil fuels (natural gas, kerosene, diesel, and propane), electricity, biogas, and solar energy.

In essence, drying has been considered as a dehydration technique which consumes high amount of energy and is responsible for significant pollutant emissions in the atmosphere.

Because the environmental impact provoked by industry in the world, specially drying systems, it is crucial to decrease energy consumption and pollutant emissions, in order to control greenhouse, so, the uses of renewable energy sources and/or combined and advanced drying techniques are strongly recommended. Thus, energy saving in industrial drying by partial recovery of heat of the exhaust air in the dryer is discussed, and the application of heat pump technology in drying process is considered. This technique has demonstrated to be more efficient, in terms of energy savings and quality product retention, especially in highly thermo-sensitive materials, such as bioproducts.

Acknowledgments The authors would like to express their thanks to CNPq (Conselho Nacional de Desenvolvimento Científico e Tecnológico, Brazil), CAPES (Coordenação de Aperfeiçoamento de Pessoal de Nível Superior, Brazil), and FINEP (Financiadora de Estudos e Projetos, Brazil) for supporting this work; to the authors of the references in this paper that helped in our understanding of this complex subject, and to the Editors by the opportunity given to present our research in this book.

References

1. Wolf, W., Spiess, W.E.L., Jung, G.: Sorption isotherms and water activity of food materials. Elsevier, New York (1985)
2. van 't Land, C.M.: Industrial drying equipment: selection and application, Marcel Dekker, Inc., New York (1991)
3. Mujumdar, A.S.: Principles, classification, and selection of dryers. In: Mujumdar, A.S. (ed.) Handbook of industrial drying, pp. 4–32. CRC Press, Boca Raton (2007)
4. Strumillo, C., Kudra, T.: Drying: principles, science and design. Gordon and Breach Science Publishers, New York (1986)
5. Keey, R.B.: Drying of loose and particulate materials. Hemisphere Publishing Corporation, New York (1992)
6. Kemp, I.C., Bahu, R.E.A.: New algorithm for dryer selection. In: 9th International Drying Symposium (Drying'94), vol. A, pp. 439–446. Gold Coast, Australia (1994)
7. Baker, C.G.J., Lababidi, H.M.S.: Application of fuzzy expert systems in dryer selection. In: 11th International Drying Symposium. (Drying'98), vol. A, pp. 448–455. Halkidiki, Greece (1998)
8. Matasov, A., Menshutina, N., Kudra, T.: Information system for the selection of dryer. In: 11th International Drying Symposium (IDS'98), vol. A, pp. 624–629. Halkidiki, Greece (1998)
9. Kemp, I.C.: Progress in dryer selection techniques. In: 11th International Drying Symposium (IDS'98), vol. A, pp. 668–675. Halkidiki, Greece (1998)
10. Lengyel, A., Kerekes, B., Sikolya, L.: Energy-demand of fruit-drying, regarding to the bonding energy of water. In: 11th International Drying Symposium (IDS'98), vol. B, pp. 1082–1092. Halkidiki, Greece (1998)
11. Muralidhara, H.S.: Advanced dewatering techniques and their impact on drying technologies. In: 8th International Drying Symposium (Drying'92), Part A, pp. 200–208. Montreal, Canada (1992)
12. Szentmarjay, T., Pallai, E., Szekrenyessy, K.: Product quality and operational parameters of drying. In: 10th International Drying Symposium (IDS'96), vol. B, pp. 839–846. Krakow, Poland (1996)
13. McMinn, W.A.M., Magee, T.R.A.: Physical characteristics of dehydrated potatoes—part II. J. Food Eng. **33**, 49–55 (1997)
14. Pappas, C., Tsami, E., Vlachopanagiotou, V., Marinos-Kouris, D.: The rehydration kinetics of new-vacuum dehydrated fruits: the role of the process conditions. In: 11th International Drying Symposium (IDS'98), vol. B, pp. 1115–1122. Halkidiki, Greece (1998)
15. Karel, M.: Optimization of quality of dehydrated foods and biomaterial. In: 8th International Drying Symposium (IDS'92), Part A, pp. 3–16. Montreal, Canada (1992)
16. Sokhansanj, S.: Drying of foodstuffs. In: Mujumdar, A.S. (ed.) Handbook of industrial drying, vol. 1, pp. 589–625. Marcel Dekker Inc, New York (1995)
17. Rizvi, S.S.H.: Thermodynamic properties of foods in dehydration. In: Rao, M.A., Rizvi, S.S. H. (eds.) Engineering properties of foods, pp. 223–309. Marcel Dekker, Inc., New York (1995)
18. Keey, R.B.: Colors development on drying. In: 14th International Drying Symposium (IDS'2004), vol. A, pp. 33–47. São Paulo, Brazil (2004)
19. Lewicki, P.P., Drzewucka-Bujak, J.: Effect of drying on tissue structure of selected fruits and vegetables. In: 11th International Drying Symposium (IDS'98), vol. B, pp. 1093–1099. Halkidiki, Greece (1998)
20. Lewicki, P.P., Piechnik, H.: Computer image analysis of shrinkage during food dehydration. In: 10th International Drying Symposium (IDS'96), vol. B, pp. 793–800. Krakow, Poland (1996)
21. Suzuki, T.; Imamura, K.; Yamamoto, K. Okazaki, M.: Stabilizing effect on freeze-dried proteins by amorphous matrices of sugar. In: 10th International Drying Symposium (IDS'96), vol. B, pp. 1261–1266. Krakow, Poland (1996)

22. Yamamoto, S.: Effects of carbohydrates on enzyme stabilization during drying. In: 10th International Drying Symposium (IDS'96), vol. B, pp. 1267–1274. Krakow, Poland (1996)
23. Luyben, K.Ch., Liou, J.K., Bruin, S.: Enzyme degradation during drying. Biotechnol. Bioeng. **24**, 533–552 (1982)
24. Liu, X.D., Strumidłło, C., Zbicinski, I.: Protection of thermo-and xero-labile materials during thermal drying. In: 10th International Drying Symposium (IDS'96), vol. B, pp. 1237–1244. Krakow, Poland (1996)
25. Labuza, T., Roe, K., Payne, C., Panda, F., Labuza, T. J., Labuza, P.S., Krusch, L.: Storage stability of dry food systems: influence of state changes during drying and storage. In: 14th International Drying Symposium (IDS'2004), vol. A, pp. 48–68. São Paulo, Brazil (2004)
26. Pikal, M.J.: Freeze dried proteins: the effect of formulation on process design and product quality. In: 10th International Drying Symposium (IDS'96), vol. B, pp. 1245–1254. Krakow, Poland (1996)
27. Kumar, C., Karim, M.A., Joardder, M.U.H.: Intermittent drying of food products: a critical review. J. Food Eng. **121**, 48–57 (2014)
28. Chua, K.J., Mujumdar, A.S., Chou, S.K.: Intermittent drying of bio-products—on overview. Bioresour. Technol. **90**, 285–295 (2003)
29. Saravacos, G.D.: Mass transfer properties of foods. In: Rao, M.A., Rizvi, S.S.H. (eds.) Engineering properties of foods, pp. 169–221. Marcel Dekker, Inc., New York (1995)
30. Jayaraman, K.S., Das Gupta, D.K.: Drying of fruits and vegetables. In: Mujumdar, A.S. (ed.) Handbook of Industrial Drying, vol. 1, pp. 643–690. Marcel Dekker, Inc., New York (1995)
31. Bimbinet, J.J., Lebert, A.: Food drying and quality interactions. In: 8th International Drying Symposium (IDS'92), Part A, pp. 42–57. Montreal, Canada (1992)
32. Bruin, S., Luyben, K. Ch.A.M.: Drying of food materials: a review of recent developments. In: Mujumdar, A.S. (ed.) Advances in Drying, vol. 1, pp. 155–215. Hemisphere Publishing Corporation, Washington (1980)
33. Menshutina, N.V., Puchkov, M.N., Goncharova, S.V., Voynovskiy, A.A.: Improvement of information system "dryinf". In: 13th International Drying Symposium (IDS'2002), vol. A, pp. 434–439. Beijing, China (2002)
34. Ogura, H., Hamaguchi, N., Kage, H., Mujumdar, A.S.: Energy and cost estimation: application of chemical heat pump dryer to industrial ceramics drying In: 13th International Drying Symposium (IDS'2002), vol. B, pp. 1105–1114. Beijing, China (2002)
35. Chua, K.J., Mujumdar, A.S., Hawlader, M.N.A., Chou, S.K., Ho, J.C.: Batch drying of banana pieces—effect of stepwise change in drying air temperature on drying kinetics and product colour. Food Res. Int. **34**(8), 721–731 (2001)
36. Strumillo, C., Jones, P. L.; Zylla, R.: Energy aspects in drying. In: Mujumdar, A.S. (ed.) Handbook of Industrial drying, pp. 285–305. CRC Press, Boca Raton (2007)
37. Raghavan, V.G.S., Rennie, T.J., Sunjka, P.S., Orsat, V., Phaphuangwittayakul, W., Terdtoon, P.: Energy aspects of novel techniques for drying biological material. In: 14th International Drying Symposium (IDS'2004), vol. B, pp. 1021–1028. São Paulo, Brazil (2004)
38. Baker, C.G.J.: Energy efficient dryer selection—an update on developments. In: 14th International Drying Symposium (IDS'2004), vol. B, pp. 957–964. São Paulo, Brazil (2004)
39. Constant, T.; Perre, P.; Moyne, C.: Microwave drying of light concrete: from transport mechanisms to explanation of energy saving. In: 8th International Drying Symposium (IDS'92), Part A, pp. 617–629. Montreal, Canada (1992)
40. Menshutina, N.V., Goncharova, S.V., Kudra, T.: Saving of energy in drying using a newly developed portal. In: 14th International Drying Symposium (IDS'2004), vol. A, pp. 672–679. São Paulo, Brazil (2004)
41. Turner, I.W., Bremhorst, K.: The effect of combined microwave and convective drying of porous media on drying kinetics. In: 8th International Drying Symposium (Drying'92), Part A, pp. 635–649. Montreal, Canada (1992)
42. Raghavan, V.G.S.: Commercial grain dryers. In: Chakraverty, A., Mujumdar, A.S., Raghavan, H.S. (eds.) Handbook of postharvest technology: cereals, fruits, vegetables, tea, and spices. Marcel Dekker, Inc., New York (2003)

43. Baker, C.G.J., McKenzie, K.A.: Energy consumption of industrial spray dryers. In: 13th International Drying Symposium (IDS'2002), vol. A, pp. 645–652. Beijing, China (2002)
44. Pettersson, M., Stenström, S.: Experimental evaluation of electric infrared dryers in the paper industry. In: 11th International Drying Symposium (IDS'98), vol. B, pp. 1587–1594. Halkidiki, Greece (1998)
45. Santos, B.M., Queiroz, M.R., Borges, T.P.F.: A solar collector design procedure for crop drying. In: 14th International Drying Symposium (IDS'2004), vol. B, pp. 1042–1048. São Paulo, Brazil (2004)
46. Kudra, T.: Instantaneous dryer indices for energy performance analyses. Chem. Process. Eng. Pol. Acad. Sci. **19**(1), 163–172 (1998)
47. Dengying, L., Na, Z., Ruixian, C.: Preliminary analysis of performance evaluation criteria for modern drying system. In: 13th International Drying Symposium (IDS'2002), vol. B, pp. 1127–1132. Beijing, China (2002)
48. Kemp, I.C.: Reducing dryer energy use by process integration and pinch analysis. In: 14th International Drying Symposium (IDS'2004), vol. B, pp. 1029–1036. São Paulo, Brazil (2004)
49. Kudra, T., Mujumdar, A.S.: Advanced drying technologies. Marcel Dekker Inc, New York (2002)
50. Berghel, J.: Increased capacity in an existing spouted dryer using a heating tube. In: CD ROM, 16th International Drying Symposium (IDS'2008) Hyderabad, India, pp. 662–666 (2008)
51. Schoenan, G.S., Arinze, E.A., Sokhansanj, S.: Simulation and optimization of energy systems for in-bin drying of canola grain (rapeseed). Energ. Conv. Manag. **36**(1), 41–59 (1995)
52. Naghavi, Z., Moheb, A., Ziaei-rad, S.: Numerical simulation of rough rice drying in a deep-bed dryer using non-equilibrium model. Energ. Conv. Manage. **51**(2), 258–264 (2010)
53. Aregba, A.W., Nadeau, J.P.: Comparison of two non-equilibrium models for static deep bed drying by numerical simulation. J. Food Eng. **78**, 1174–1187 (2007)
54. Jumah, R.Y., Mujumdar, A.S.: Dryer emission control systems. In: Mujumdar, A.S. (ed.) Handbook of Industrial Drying, vol. 2, pp. 1179–1226. Marcel Dekker, Inc., New York (1995)
55. Strumillo, C., Jones, P.L., Zylla, R.: Energy aspects in drying. In: Mujumdar, A.S. (ed.) Handbook Industrial Drying, vol. 2, pp. 1241–1275. Marcel Dekker, Inc., New York (1995)
56. Kudra, T., Mujumdar, A.S.: Special drying techniques and novel dryers. In: Mujumdar, A.S. (ed.) Handbook of Industrial Drying, vol. 2, pp. 1087–1149. Marcel Dekker Inc., New York (1995)
57. Kudra,T., Mujumdar, A.S.: Advanced drying technologies. 2 edn. CRC Press, Boca Raton (2009)
58. Sztabert, Z., Kudra, T.: Cost estimation methods for drying. In: Mujumdar, A.S. (ed.) Handbook of Industrial Drying. vol. 2, pp. 1227–1240. Marcel Dekker Inc., New York (1995)
59. Dimitriadis, A.N., Akritidis, C.B., Arvanitis, K.G.: Energy saving in drying plants by partial recovery of the latent heat. In: CD ROM, 12th International Drying Symposium (IDS'2000). Noordwijkerhout, Netherlands, Paper 289 (2000)
60. Moraitis, C.S., Akritidis, C.B.: Energy saving in industrial drying plants by partial recovery of the latent heat of exhaust air. In: 10th International Drying Symposium (IDS'96), vol. A, pp. 489–496. Krakow, Poland (1996)
61. Yongdong, Z.: Research development of wood drying energy conservation in China. In: CD ROM, 16th International Drying Symposium (IDS'2008), pp. 1138–1141. Hyderabad, India (2008)
62. Ogura, H., Kage, H., Matsuno, Y., Mujumdar, A.S.: Studies on a novel chemical heat pump dryer using a gas-solid reaction. In: CD ROM, 12th International Drying Symposium (IDS'2000), Noordwijkerhout, Netherlands, Paper 14 (2000)
63. Prasertsan, S.; Saen-Saby, P.; Prateepchaikul, G.; Ngamsritrakul, P. Effects of product drying rate and ambient condition on the operating modes of heat pump dryer. In: 10th International Drying Symposium (IDS'96), vol. A, pp. 529–534. Krakow, Poland (1996)

64. Poomsa-ad, N., Atthajariyakul, S., Wiset, L., Tipsanprom, W.: Performance of heat pump dryer for nam-dok-mai mango under different medias. In: CD ROM, 16th International Drying Symposium (IDS'2008), pp. 1614–1617. Hyderabad, India (2008)
65. Kaiser, S., Tsotsas, E.: Avoidance of fouling from dryer exhaust gases as a presupposition for efficient energy recovery—an experimental approach. In: CD ROM, 12th International Drying Symposium (IDS'2000). Noordwijkerhout, Netherlands, Paper 55 (2000)

Intermittent Drying: Fundamentals, Modeling and Applications

A.G. Barbosa de Lima, J.M.P.Q. Delgado, S.R.F. Neto and C.M.R. Franco

Abstract This chapter focuses on the intermittent drying of wet porous bodies. Drying process was simulated assuming liquid diffusion as the sole mass transport mechanism and constant mass diffusion coefficient. Application has been done to ellipsoidal solids. The transient mass diffusion equation in a prolate spheroidal coordinate system was used to study the process in two-dimensional cases. Results are presented, changing the dimensionless tempering time, aspect ratio of the body and number of drying passes, by using the continuous drying process, drying rate, energy cost, drying time and quality of the product post-drying as comparison parameters.

Keywords Tempering · Simulation · Mass · Quality · Ellipsoid

A.G. Barbosa de Lima (✉)
Department of Mechanical Engineering, Federal University of Campina Grande,
Av. Aprígio Veloso, 882, Bodocongó, Campina Grande, PB 58429-900, Brazil
e-mail: antonio.gilson@ufcg.edu.br

J.M.P.Q. Delgado
CONSTRUCT-LFC, Faculty of Engineering (FEUP), University of Porto,
Porto, Portugal
e-mail: jdelgado@fe.up.pt

S.R.F. Neto
Department of Chemical Engineering, Federal University of Campina Grande (UFCG),
Campina Grande, PB 58429-900, Brazil
e-mail: fariasn@deq.ufcg.edu.br

C.M.R. Franco
Department of Physics and Mathematics, CES, Federal University of Campina Grande
(UFCG), Cuité, PB 58175-000, Brazil
e-mail: celiarufino@ufcg.edu.br

J.M.P.Q. Delgado and A.G. Barbosa de Lima (eds.),
Drying and Energy Technologies, Advanced Structured Materials 63,
DOI 10.1007/978-3-319-19767-8_2

1 Basic Concepts in Drying

Hygroscopic materials undergo changes in its water content according to the conditions of the ambient air surrounding it. In the case of agricultural products, control the moisture content is crucial for the preservation of their quality and to avoid waste. In this context, the drying is one of the most widely used methods for this purpose.

Since antiquity, dehydration has been used in food preservation and grain drying. Technical processes were passed from generation to generation through the knowledge acquired with the ancestors, and gradually improved. Dehydration or drying operations are important processes in the chemical and food industries, as well as the storage of grain and other biological products. Reduce the moisture content of agricultural products to a safe level for storage is an usual practical postharvest.

Drying is a process involving heat and mass transfer (moisture) between a hygroscopic product and drying air. In the drying process of a wet porous solid, the air supply heat to the solid and absorbs water of it in vapor phase. In general, drying is conducted with higher air temperatures than the boiling point of the liquid to be eliminated. The heating of the drying air, in order, to reduce the relative humidity and increase its enthalpy and evaporative capacity, must be properly controlled to avoid physical, chemical and biological damage that can cause to the product.

The water vapor present into the product tends to occupy all intercellular space available generating pressure, including the interface between the product and the air surrounding it. This pressure is called the partial pressure of water vapor at the surface (P_s) of product. On the other hand, water vapor in air performs a partial pressure so called as partial pressure of water vapor in the air (P_{air}).

If the moisture content of the product is very low, so that the partial pressure of water vapor at the surface is lower than the partial pressure of water vapor in the ambient air, the product will gain moisture, which may cause the emergence of fungi and damage it. In the drying process, the moisture removal is achieved by water movement, due to a difference in partial pressure of water vapor between the surface of the product to be dried and the air that surrounds it. Then, drying of the product will occur if ($P_s > P_{air}$) (see Fig. 1).

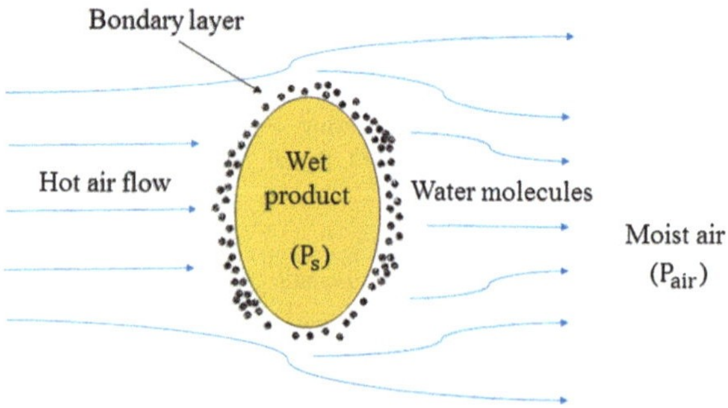

Fig. 1 Representation of the movement of water during drying

Unsaturated air in contact with product causes evaporation of water peripheral which generates a moisture gradient in the layers of the product. Through diffusion, the internal water moves to the periphery, where evaporates, creating new moisture gradients, making the process to continue. In drying with heated air, beyond of the peripheral evaporation, the temperature increase causing an increase in the internal pressure, creating pressure gradients that are added with the water gradient effect. Then, diffusion occurs in presence of the two effects and progress with high intensity. The heated air and with low relative humidity has its evaporative capacity and energy exchange increased by increasing its enthalpy. When the heating happens in the product also increases liquid evaporation and internal pressures [1].

Hygroscopic products during drying have moisture contents commonly defined as: initial, critical and equilibrium moisture content. The initial moisture content is the amount of water into the product when starting the drying process. This water can be strongly bounded to the dry mass of the product, which is difficult to be removed, or in the form of free water which can be removed easily depending on the conditions of the environment in which the product is placed. The moisture content is which occurs changes in the drying rate from constant for decreasing is called critical moisture content. The equilibrium moisture content is reached when the wet product is in equilibrium with the drying air, for a given thermodynamic state of the air. In this case, there is no moisture flux between them. According to Fioreze [2], this does not mean that the product moisture and air are equal, but that the vapor pressures at the product surface and air are equal. When this equilibrium is reached, the moisture inside the product becomes almost uniform.

Drying process presents three classical steps as pictured in the Fig. 2. Early of the drying occurs rising of the temperature of the product and pressure of water vapor (warm-up period). At this moment, the drying rate $d\overline{M}/dt$ is increasing, due to the relative humidity in the boundary layer is less than 100 %, happening what is usually called accommodation period [3]. The increases in drying rate continue to

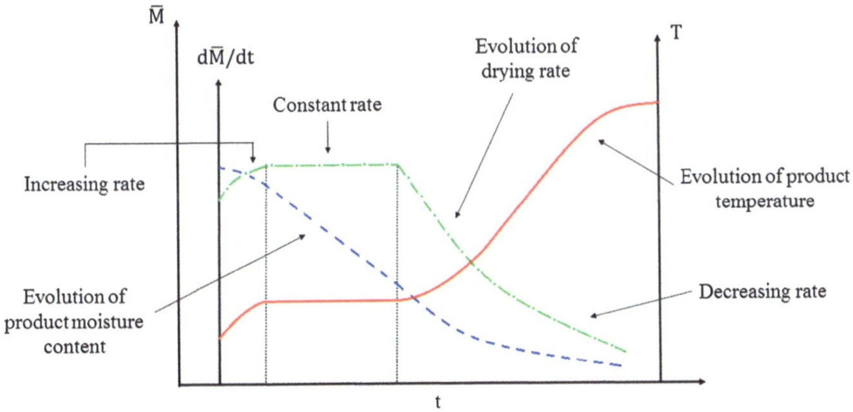

Fig. 2 Typical drying curve

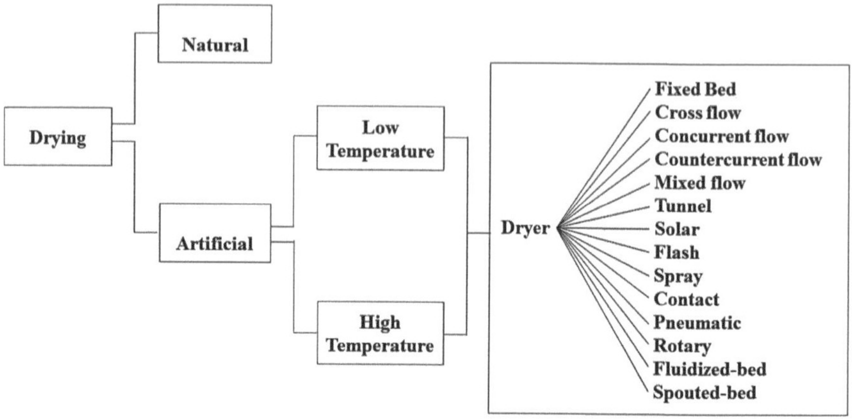

Fig. 3 Drying methods and dryer types

happen until the point where equivalence between heat and mass (water) transfer, occurs that is, the constant rate period. At the start of drying, the product is completely wet and the water flows in liquid phase under a hydraulic gradient, and under this condition, the product temperature equals the wet bulb temperature. While there is surface moisture to progress the evaporation process, the drying rate remains constant. The ending point of the constant rate period is called the point of critical moisture (moisture content at which water ceases to behave as free water).

As drying proceeds, and had passed of the point of critical moisture, the moisture content continue to decrease and the water in the liquid phase forms liquid bridges inside the porous solid. From this point begins the period of decreasing drying rate. During this period, occurs a reduction in moisture migration from the interior to the surface of the product and the heat transfer is not equivalent to the mass transfer. The product temperature exceeds the wet bulb temperature until to the drying air temperature or dry bulb temperature. Finally, the drying occurs within the product until that equilibrium moisture content is reached, i.e., when the amount of evaporated water equals the amount condensed water at the surface of the solid [1, 4, 5].

Drying can be realized of natural or artificial manner (Fig. 3). In natural drying, air movement is made by the wind and the energy for moisture evaporation comes of air drying potential and direct incidence of sun energy. Artificial drying is characterized by the use of manual or mechanical processes both in product handling and in air passage through the product. In drying with forced ventilation we can employ low or high temperature, combined drying, and other techniques. Details about this topic can be found at [1, 2, 4–9].

On operation, the dryers can be classified in two types [1, 2, 4, 5]:

(a) **Continuous dryers** The product is constantly under the action of heat, until that its moisture content reaches a desired value. For this, the product passes through flow regulation mechanism that will determine the time of exposure to drying air, also called residence time.

(b) **Intermittent dryers** Product passes several times through the dryer prior to complete drying. Thus, the product undergoes the action of heat during short time intervals, interspersed with rest periods, i.e., the product do not come into contact with the heated air during this period. In general, these dryers consist of two drying towers and a tank placed above these columns.

During drying, there is a difference between moisture located at the surface and inside the product. The surface that is in direct contact with the air tends to dry more than the central part. Thus, the product undergoes transformations and therefore, several studies have been made in order to determine the best way to accomplish such a process. The analysis in this case is made from experiments and/or simulations using a given mathematical model.

Severe drying with low air relative humidity and high air temperature may damage the product and make it unsuitable for consumption and processing. For example, when the grain loses water it has reduced its size by external compression; this effect increases as it dries. When the grain is heated increases its internal pressure and the more central layers of the grain tend to expand. The grain surface has no elastic plasticity and capability to withstand very high hydro-thermo-mechanical stresses and in this case can suffer cracking or even breakage. As the more unbalanced are the evaporation and diffusion phenomena, the greater the damage.

Thus, it can be said that the main advantage of drying with heated air is in the decreasing in drying time and the disadvantages are the cost of energy required to heat the air and the damage that it can cause to the product due to higher temperature and lower relative humidity mainly in heat-sensitive materials, such as: fruit, vegetables and grains.

Various drying techniques are reported in literature to improve the final product quality and reduce consumption of energy used. First, however, we will present the basic concepts of drying that are important for a better understanding of the physical phenomena involved in the process. Now, we discuss the technique of intermittent drying presenting its main features and a mathematical model to simulate this process.

2 Intermittent Drying

2.1 *Fundamentals, Energy Efficiency and Product Quality Aspects*

In present day, the concept of multi-stage drying has been well accepted for researchers. Several drying techniques have been proposed in order to rationalize the use of energy, as well as to reduce other problems during the drying process, such as crack, fissure, loss of germination power and vigor of the grain and seeds, non-enzymatic browning of fruits and vegetables. One of these techniques is the intermittent drying.

The intermittent drying is a process that alternates drying periods with of rest or relaxation periods. The product is subjected to the action of hot air in drying chamber at regular time intervals, i.e., heat is supplied discontinuously. In the rest period (non drying, tempering), the product goes through part of the system which does not get heated air, allowing homogenization of moisture and cooling [10, 11]. Each rest period between two other continuous drying is called pass [12]. The aim for determination of the duration and number of cycles (or tempering) in the intermittent drying process is to minimize costs (energy) and get a better final product quality [13–15]. Applies the intermittent drying only in period when the drying rate is decreasing [13, 16].

For certain materials, much of the resistance to drying process lies inside it, so, maintain certain drying air conditions applied only the surface of the material results in high gradients at the product surface, causing possible damage (cracks and deformations) and reducing process efficiency. In these cases, intermittent drying is recommended.

The intermittent drying technique is commonly used for the drying of heat-sensitive material such as: grains (wheat, soya, rice, corn, coffee), vegetables (potato), fruits (guava, banana, mango) and various kinds of herbs [16–25] and non food material such as ceramic and wood [26–30], and provides advantages over the continuous drying. In this method, the speed and uniformity of drying are the most relevant characteristics. The amount of water removed by unit drying time is considerably higher than when the drying is restarted. During the rest period (tempering period), due to moisture gradient established inside the product, there is moisture migration from the interior to the surface, until the moisture in the entire product is almost uniform. This moisture redistribution, beyond of facilitate drying when the heat application is restarted, reduces the water and temperature gradients and, as result, reduces thermal and hydric stresses and physical damage (cracks) to the product, not violating the material structure [31]. This phenomenon is namely cited as the "refreshing effect" by Nishiyama et al. [32].

Due to intermittency, it is possible to use the temperature of the heated air reaching up to 70–80 °C, without causing excessive heating of the product, in general, do not reach temperatures above from 40 to 43 °C [33, 34].

According to Shei and Chen [35] less than 2 % (dry basis) of the moisture is removed from the product after each rest period as using re-circulating rice dryer.

The effect of tempering time on the amount of moisture removed after drying, during the cooling of shelled corn, was studied by Sabbah et al. [36] and Tolaba et al. [37] and the tempering of rice by Steffe et al. [38], Walker and Bakker-Arkema [39], Fioreze [40] and Elbert et al. [41].

Knowledge of the ideal tempering time is very important because of the reduction in energy utilized in the drying to avoid over heating of the product, and mainly, the preservation of product quality. According to Steffe and Singh [42], if the tempering time is too short, cracking may occur and the quality of the grain can be affected. But with the use of a short tempering time, some advantage, such as the minimization of damage produced by chemical changes and insect and microbial activity, as well an increase in drying capacity and drier flexibility, do exist because total drying time is reduced.

Chua et al. [13] show some ways to implement the intermittent drying process:

(a) Intermittent drying whereby heat flux is supplied intermittently rather than continuously. This can be done by interrupting the air flow to provide the material a "rest" or "tempering" period, by a continuous air flow periodically heated, or by periodic variation of air flow or by combination of both;
(b) Aeration which is a drying process involving a combination of high temperature short drying period, tempering, and slow cooling followed by drying;
(c) Air reversal drying which is reversing the direction of the airflow for a period of time and then returns it to its original direction. This procedure is applied to deep bed drying of particulates to minimize temperature and moisture gradients inside the bed;
(d) Cyclic drying which is a drying process whereby the temperature, humidity or velocity of the air undergoes a specified cyclic pattern variation such as sinusoidal, square-wave or saw-tooth patterns. The operating pressure can also be cycled.

Experimental evidence of the potential benefits of intermittent temperature variation on product quality and energy saving has been demonstrated by many researchers [20, 30, 32, 43, 44]. Thus, this technique has been realized in different dryers, such as batch, fluidized bed, spouted bed, microwave and heat pump dryers [33, 45–48].

Chua et al. [49] have demonstrated that employing intermittent drying air temperature could reduce the overall color change of potato, guava and banana samples by 87, 75 and 67 %, respectively. Improvement in retention of ascorbic acid up to 20 % and β-carotene were verified by Chua et al. [49] and Chua et al. [13], respectively.

Nishiyama et al. [32] have demonstrated that the tempering temperature had a significant effect on the rate of internal moisture equilibration and the drying rate was improved subsequent to tempering, showing the benefit of allowing moisture redistribution for wheat and rough rice. Further, they author have concluded that long-grain rough rice required more tempering than short-grain rough rice.

The results of the drying studies presented for Kowalski and Pawłowski [30] allows to state that the intermittent drying can be recommended above all to drying of materials, which have a tendency to cracking during drying as, for example, ceramics and wood. Through changes of drying conditions in the pre-established instants one can avoid material fracture and thus preserve a good quality of dried products. The authors have stated that intermittent drying positively influences the quality of the dried materials without significant extension of the drying time. The best quality of the tested samples was achieved, unfortunately, in the most energy consuming process, that is, during intermittent drying with the variable air humidity. The intermittent drying with variable air temperature gives better product quality than drying in stationary conditions at similar drying times and, what is very important, by the lowest energy consumption in all the tested processes. The best product quality obtained in intermittent drying with variable air humidity and the lowest energy consumption in intermittent drying with variable air temperature suggests the advantageous combination of both these drying techniques to optimize the drying process with respect to the energy consumption and the quality of dried products. Thus, the combination of those techniques would increase the effectiveness of the drying process.

Mabrouk et al. [43] studied intermittent drying of the apple thin slice. Variations in the air temperature and air velocity values permitted to find the most economic velocities for the drying process. The numerical simulations in the intermittent tests, give more information about the behavior of the product during the periods of interruption. This study shows the considerable advantages of the intermittent drying of agricultural products and confirms the preservation of their qualities, the reduction in the drying time and in energy consumption.

Silva et al. [44] has studied the continuous and discontinuous pear drying. Experimental results were compared with the Fick's diffusion model assuming a convective boundary condition enabling the determination of an effective diffusion coefficient and convective mass transfer coefficient at the surface of the product. A good agreement was observed between modeling and experimental data. In the intermittent drying the hot airflow was turned off during night periods, for 13.5 h with velocity and temperature of the air flow 1.2 m/s and 40 °C, respectively. According to the authors, total duration of the pauses (during which the drying installation was turned off, and consequently no energy was spent), was in about 50 % of total test duration. This result points out towards a possible energy saving strategy in the intermittent process. In particular, it is demonstrated that an increase of the number of pauses can lead to significant energy savings and that drying with solar energy, which is inherently discontinuous is an effective method for this type of process.

2.2 Mathematical Modelling

Prediction of continuous and intermittent drying by the reliable numerical and analytical approaches is very important to help in the design, energy consumption computation of dryer, and to predict drying rate and moisture and temperature gradients inside the product, and to maintain quality during and post-drying.

To model continuous and intermittent drying process in solid with prolate spheroidal shape, the following assumptions were considered:

- The solid is homogeneous, isotropic and composed of water in liquid phase and solid material;
- The continuous drying process occurs under falling rate;
- The heat conduction through the prolate spheroid is neglected;
- The moisture content profile is symmetric around z-axis;
- The thermophysical properties are constants during the drying process;

- The process occurs under convective condition at the surface of the solid;
- Shrinkage of the solid is neglected.

Based on these assumptions, the following models were developed to simulate the intermittent and continuous drying process.

(a) **Continuous drying modeling**

In many drying problems the internal resistance of the solid to mass transfer is greater than its resistance to heat transfer. In these cases moisture diffusion is the mechanism that controls the process during the falling drying rate period, and the pure diffusion model can be used to predict drying process. During the falling rate period, the drying rate decays with time and continuous heating has little effect on moisture removal so, methods to improve the utilization of energy during the drying process in this period, are very important and must be studied.

The Fick's second law has been used to predict moisture content distribution inside the solid during drying process. It is given as follows:

$$\frac{\partial M}{\partial t} = \nabla \cdot (D\nabla M) \quad (1)$$

where M is the moisture content (dry basis) and D is the effective mass diffusion coefficient of the material.

To predict the diffusion phenomenon in prolate spheroids, it is necessary to write Eq. (1) for an appropriate coordinate system, in this case, in prolate spheroidal coordinates. Figure 4 shows a body with prolate spheroidal geometry.

The relationships between the Cartesian (x, y, z) and prolate spheroidal coordinate systems (μ, φ, ω) are given by Stratton et al. [50], Flammer [51] and Abramowitz and Stegun [52]:

$$x = L\sqrt{(1-\xi^2)(\eta^2-1)}\,\zeta \quad (2a)$$

$$y = L\sqrt{(1-\xi^2)(\eta^2-1)}\sqrt{(1-\zeta^2)} \quad (2b)$$

$$z = L\,\xi\,\eta \quad (2c)$$

where L_1 and L_2 are the solids dimensions, $L = \left(L_2^2 - L_1^2\right)^{1/2}$, $\xi = \cosh\mu$, $\eta = \cos\phi$ and $\zeta = \cos\omega$. The intervals of the variables ξ, η e ζ (in terms of ω) are: $0 \le \xi \le L_2/L$; $0 \le \eta \le 1$; $0 \le \omega \le 2\pi$.

By calculating the metric coefficients and the Laplacian in the new coordinate system, it can be obtained the following 3-D transient diffusion equation:

$$\frac{\partial M}{\partial t}=\left[\frac{1}{L^2(\xi^2-\eta^2)}\frac{\partial}{\partial\xi}\left((\xi^2-1)D\frac{\partial M}{\partial\xi}\right)\right]+\left[\frac{1}{L^2(\xi^2-\eta^2)}\frac{\partial}{\partial\eta}\left((1-\eta^2)D\frac{\partial M}{\partial\eta}\right)\right]$$
$$+\left[\frac{\sqrt{1-\varsigma^2}}{L^2(\xi^2-1)(1-\eta^2)}\frac{\partial}{\partial\varsigma}\left(\left(\sqrt{1-\varsigma^2}\right)D\frac{\partial M}{\partial\varsigma}\right)\right] \quad (3)$$

Using the symmetry of the body around the z-axis, Eq. (3) can be written as follows:

$$\frac{\partial M}{\partial t}=\left[\frac{1}{L^2(\xi^2-\eta^2)}\frac{\partial}{\partial\xi}\left(D(\xi^2-1)\frac{\partial M}{\partial\xi}\right)\right]$$
$$+\left[\frac{1}{L^2(\xi^2-\eta^2)}\frac{\partial}{\partial\eta}\left(D(1-\eta^2)\frac{\partial M}{\partial\eta}\right)\right] \quad (4)$$

The following boundary conditions during continuous drying periods are used:

- Free surface: Diffusive mass transfer is equal to convective mass transfer at the surface of the solid.

$$-\frac{D}{L}\sqrt{\frac{(\xi^2-1)}{(\xi^2-\eta^2)}}\frac{\partial M}{\partial\xi}\bigg|_{\xi=\xi_f}=h_m[M(\xi=\xi_f,\eta,t)-M_e], \quad (5)$$
$$\xi_f=L_2/L \text{ at the surface}$$

- Planes of symmetry: the angular and radial gradients of moisture content are equal to zero at the planes of symmetry. Then, we can write:

$$\frac{\partial M(\xi;1;t)}{\partial\eta}=0 \quad (6)$$

$$\frac{\partial M(\xi;0;t)}{\partial\eta}=0 \quad (7)$$

$$\frac{\partial M(1;\eta;t)}{\partial\xi}=0 \quad (8)$$

- Initial conditions inside solid: the moisture content is constant and uniform.

$$M(\xi;\eta;t=0)=M_o \quad (9)$$

Fig. 4 Characteristics of a prolate spheroid

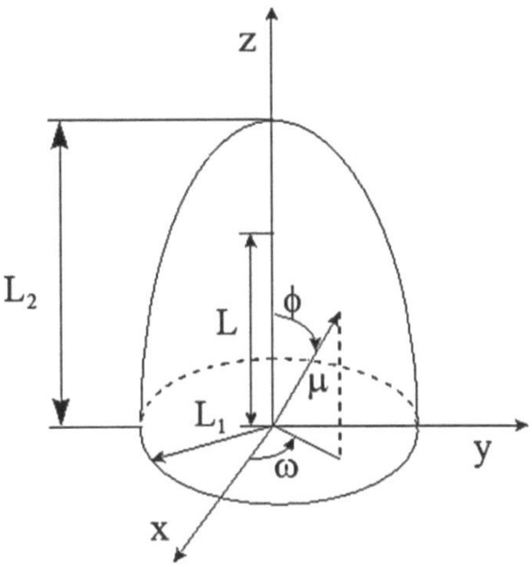

From the solution of Eq. (4) it can be calculated the average moisture content of the body during drying process as follows [53]:

$$\bar{M} = \frac{1}{V}\int_{V} M dV \tag{10}$$

where

$$dV = \frac{L^3(\xi^2 - \eta^2)}{\sqrt{(1-\zeta^2)}} d\xi d\eta d\zeta \tag{11}$$

is the infinitesimal volume within the solid. Integration of the Eq. (11) over the solid (Fig. 4) gives as result:

$$V = \frac{2}{3}\pi L_1^2 L_2 \tag{12}$$

(b) **Tempering model (Intermittent drying)**

The main difference between the continuous and intermittent drying (tempering) processes is in the initial and boundary conditions used in the model. During tempering, the moisture content profile is calculated through time, assuming the surface of the solid to be impermeable. This last assumption requires the development of a new mathematical modeling to describe mass transfer inside the solid. Under these new conditions, variations in the average moisture content of the solid are negligible during rest period, and only a change in the moisture distribution inside the solid occurs. During tempering period, the moisture content inside the product will be equalized through moisture diffusion. The ideal tempering time can be obtained by considering a final flat moisture distribution. To simulate intermittent drying, the moisture content profile inside the solid at the end of the continuous drying period must be known. Thus, the mathematical modeling that describes the tempering process is given by Eqs. (4) and (6) and the new following initial and boundary conditions:

$$\frac{\partial M(L_2/L; \eta; t)}{\partial \xi} = 0 \tag{13}$$

$$M(\xi; \eta; t = 0) = f(\xi; \eta) \tag{14}$$

Function $f(\xi; \eta)$ in the initial period of tempering corresponds to the moisture content profile within the solid at the end of the continuous drying period. It obeys variation with the radial and angular coordinates. Because of the impermeability condition, the drying rate in the solid must be equal to zero. In this case, the rate of moisture storage in the solid must be equal to the change in moisture content inside it, during tempering period. Numerically, the condition given by Eq. (13) corresponds to the situation in which the moisture content at the nodal points on the surface is equal to this value at the nodal points immediately near to the surface. From a mathematical standpoint, this is the same treatment as that used under the symmetry condition. Then, we can express this condition as follows:

$$\left(\frac{\partial M}{\partial t}\right)_P dV = \frac{D}{L}\left(\sqrt{\frac{\xi_s^2 - 1}{\xi_s^2 - \eta_P^2}}\right)\frac{\partial M}{\partial \xi}\bigg|_s dS_\xi \tag{15}$$

being

$$dS_\xi = \frac{L^2\sqrt{(\xi^2 - \eta^2)}\sqrt{(\xi^2 - 1)}}{\sqrt{(1 - \zeta^2)}} d\eta d\zeta \tag{16}$$

The parameter dS_ξ represent mass flow area of an infinitesimal element in the prolate spheroidal coordinate system, as indicated in Fig. 4.

2.3 Numerical Solution Procedure

Several numerical methods (finite-difference, finite-element, finite-volume, etc.) are used to solve the physical problem of mass transfer during drying process in solid with different geometries. Herein, it was used the finite-volume method. The numerical solution of the problem utilizing the finite-volume technique is obtained by integrating Eq. (4) with over a volume and time. For a fully implicit formulation and practice B, the discretized equation is given by [54–56]:

$$A_P M_P = A_E M_E + A_W M_W + A_N M_N + A_S M_S + A_P^0 M_P^0 \tag{17}$$

Application of the Eq. (17) over the computational domain derives a system of linear algebraic equation, which was solved using the Gauss-Seidel iterative algorithm. A 20 × 20 central-volume grid was used in the simulations.

The calculation started with the given initial condition and stopped when the following convergence criterion was satisfied at each point of the computational domain:

$$\left|M^{k-1} - M^{k}\right| \leq 10^{-8} \tag{18}$$

where k represent k-th iteration.

More details about numerical procedure can be found in the reported literature [12, 57–62].

To determine the tempering Fourier required equalizing the moisture content inside the solid, it is necessary to adopt a stop criterion, which is dependent on the accuracy of the model; in this work $\left|M^{*^{o}} - M^{*}\right| \leq 10^{-7}$ was adopted as the stop criterion at all the nodal points. The superscript o represents the old time and M^* is the dimensionless moisture content.

3 General Application

Herein, results are presented in term of dimensionless variables moisture content, time, and angular and radial coordinates, as follows:

$$M^* = \frac{M - M_e}{M_o - M_e} \tag{19}$$

$$\eta^* = \eta \tag{20}$$

$$\xi^* = \xi \tag{21}$$

$$t_m^* = \frac{Dt}{L_1^2} \tag{22}$$

The ideal tempering time was obtained for spheroidal bodies with aspect ratios L_2/L_1 = 1.1, 2.0 and 5.0 for an initial dimensionless tempering time corresponding to $t_m^* = Dt/L_1^2 = 0.01098$. For L_2/L_1 = 1.1, 2.0 and 5.0 the ideal tempering Fourier were 0.42700, was 0.73200 and 0.79300, respectively. Differences in tempering

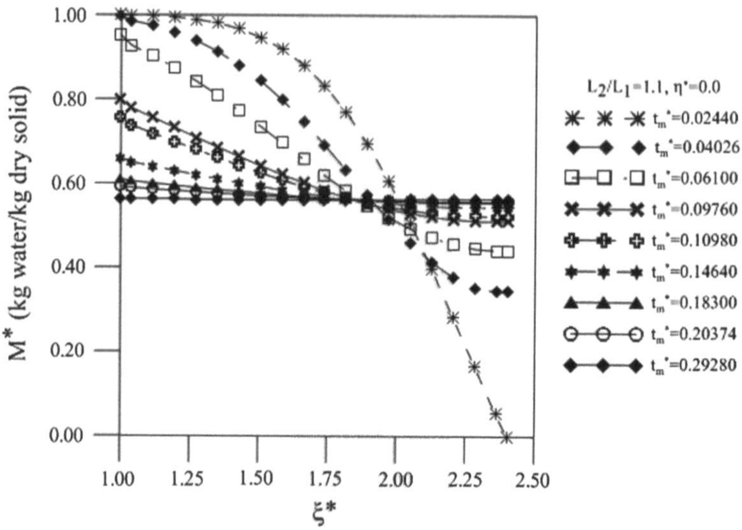

Fig. 5 Radial distribution of the moisture content inside a prolate spheroidal solid (L_2/L_1 = 1.1) during tempering period (r* = 0.0)

time among these cases, mainly between the first and the second cases, are due to the dimensions of the bodies and the moisture content gradients inside the solid at the beginning of the tempering period.

Figure 5 illustrates the moisture content inside the prolate spheroidal solid as a function of the dimensionless radial coordinate, during the tempering period, for aspect ratio L_2/L_1 = 1.1. In this case, tempering period began at $t^*_m = 0.02440$. It can be see that by increasing the dimensionless tempering time, the moisture content profile asymptotically approximated to the average value established at the begin of the tempering period. Variations in the moisture content are more intense in the first instants of drying.

Figure 6 shows the behavior of the moisture content at the center and solid surface of the body (z = 0; y = L_1) during the tempering period. The tempering period started at $t^*_m = 0.02440$. Since loss of moisture at the surface does not occur, the moisture content at this point increases with tempering time due to the decrease in moisture content in the central region of the body. This process will increase the drying rate during the post-tempering period because the difference between moisture content at the surface of the solid and equilibrium moisture content is increased.

Using the tempering mathematical model, two cases were analyzed: (a) variation in the dimensionless tempering time, and (b) number of pass of drying. Figure 7 shows the effect of the tempering time on the drying rate for a prolate spheroid with

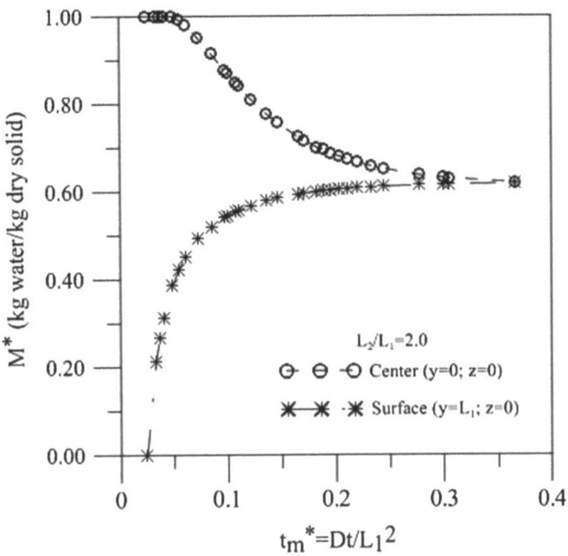

Fig. 6 Moisture content as a function of the dimensionless time for a prolate spheroidal solid with the aspect ratio $L_2/L_1 = 2.0$ during the tempering period

aspect ratio $L_2/L_1 = 2.0$. Tempering started at $t_m^* = 0.01098$ and one pass of drying was used. As compared with continuous drying process, the drying rate increased by increasing the dimensionless tempering time. The drying rate slightly changed as aspect ratio has changed from 1.1 to 5.0. The curves show that the final average moisture content was approximately equal in all cases for the same aspect ratio.

Figure 8 present plots of the average moisture content versus dimensionless time for aspect ratio $L_2/L_1 = 2.0$. In this case the start of tempering is longer than the former case. We can see that occurred an increase in drying rate, greater than that the former cases. Comparing the cases in Figs. 7 and 8, it can be observed that the drying rate increased with the increases in tempering period. This behavior has occurred for increased aspect ratio too. This effect is greater for smaller average moisture content, ranging from 0.40 to 0.10. In all cases, the drying rate is larger than that found in continuous drying, thus energy saving and improving product quality due to reduced stress inside the product is obtained.

Figure 9 shows the effect of an increase in dimensionless tempering time and the use of two passes of drying on the drying rate of prolate spheroid bodies with aspect ratio $L_2/L_1 = 2.0$. It can be seen that an increase in drying rate exists, as compared with the continuous drying process. As tempering time increases, both total amount and drying rate are increased. Thus, the dimensionless time required to complete drying of the wet porous solid is reduced. Practically no effect on drying rate is seen for dimensionless average moisture content lower than 0.15 kg/kg dry basis.

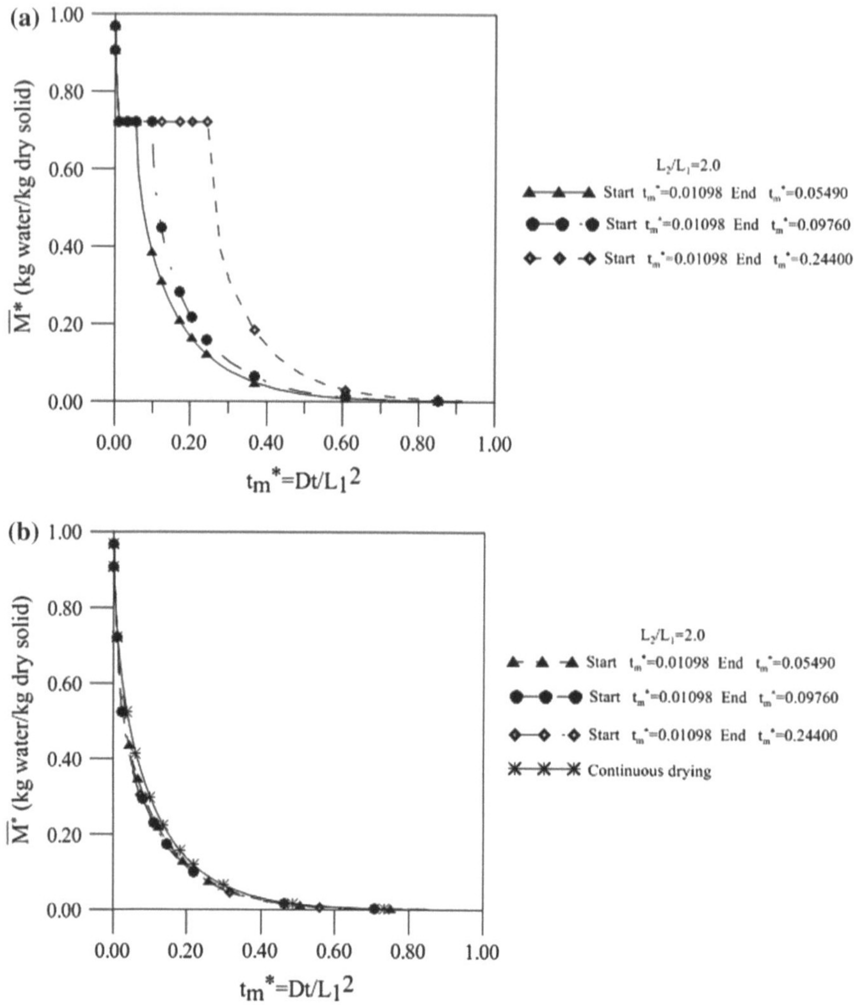

Fig. 7 Dimensionless average moisture content as a function of dimensionless time for a prolate spheroid with aspect ratio 2.0 and different tempering period (one-pass drying). Tempering period has started in 0.01098. **a** Including tempering period, **b** excluding tempering period

Figure 10 illustrates the effect of the use of multipass drying on the drying rate for case $L_2/L_1 = 2.0$, as a function of dimensionless time. The analysis of the curves shows that drying in four passes increases the drying rate more than other cases. However, the use of two passes of drying presented better results, mainly in the

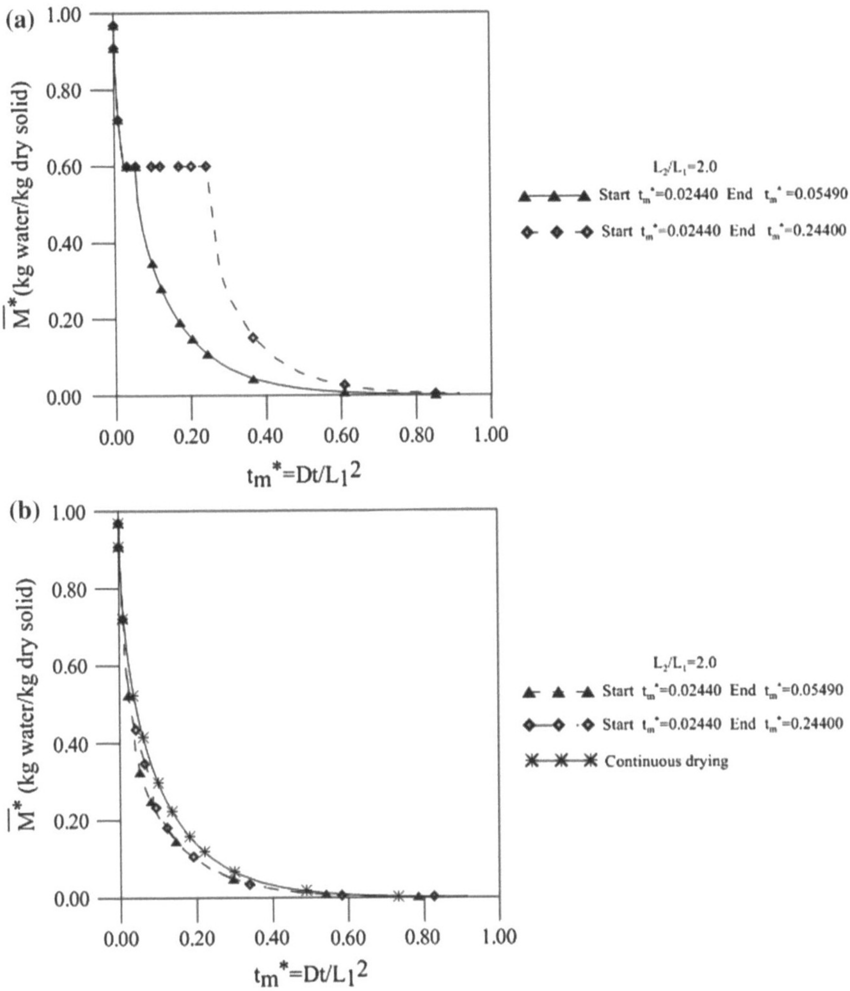

Fig. 8 Dimensionless average moisture content as a function of dimensionless time for a prolate spheroid with aspect ratio 2.0 and different tempering period (one-pass drying). Tempering period has started in 0.02440. **a** Including tempering period, **b** excluding tempering period

second stage of drying. Having a reduced drying time with a lower drying rate is very important in increasing the energy efficiency of the dryer and in reducing internal thermo-hydro-mechanical stresses in the dried material. The multipass drying, with four passes or more can be useful when the product is very sensitive to high temperatures.

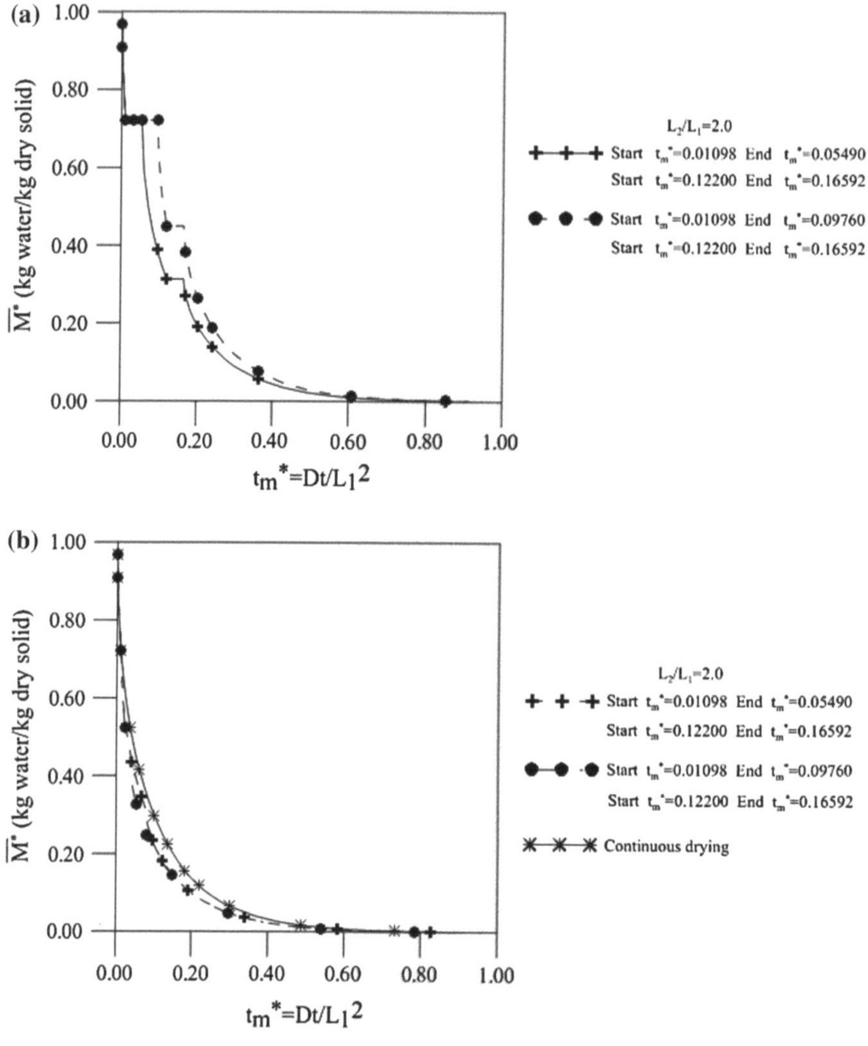

Fig. 9 Dimensionless average moisture content as a function of dimensionless time for a prolate spheroid with aspect ratio 2.0 and different tempering period (two-pass drying). Tempering period has started in 0.02440. **a** Including dimensionless tempering period, **b** excluding dimensionless tempering period

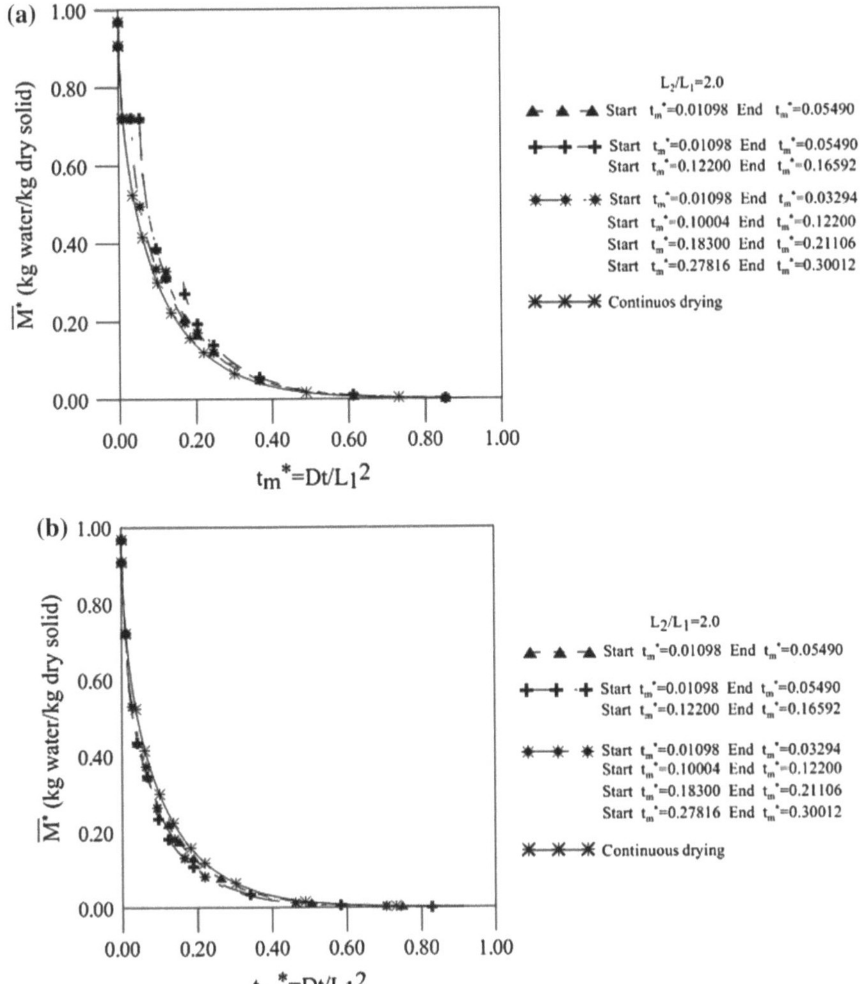

Fig. 10 Predicted drying curves for multi-pass and continuous drying of a prolate spheroidal solid with aspect ratio 2.0. **a** Including dimensionless tempering period, **b** excluding dimensionless tempering period

4 Concluding Remarks

In this chapter, a mathematical modeling to predict continuous and intermittent moisture migration in bodies with prolate spheroidal shape was developed. Numerical procedure using the finite-volume technique was applied to solve the governing equation.

Analysis of the results permits us to conclude that the intermittent drying process (tempering) is an important method to minimize energy utilization in the drying process, and it can be applied in industrial dryers to increase drying efficiency, to reduce energy costs, and maintaining product quality post-drying. In all cases studied, the drying rate after tempering was higher than the in a continuous process. The drying rate of prolate spheroids in processes with two tempering periods was higher than the other cases studied. It was verified that, both, the instant in which tempering period has started and the period of tempering affect the drying rate of the product after tempering.

Further, this work shows that the methodology presented herein has great potential and can be applied for different spheroidal shape ranging from sphere to infinite cylinder by changing the aspect ratio only. Variations in the unsteady-state diffusion equation are unnecessary.

Acknowledgments The authors would like to express their thanks to Brazilian Research agencies CNPq (Conselho Nacional de Desenvolvimento Científico e Tecnológico) and CAPES (Coordenação de Aperfeiçoamento de Pessoal de Nível Superior) for supporting this work, and are also grateful to the authors of the references in this chapter that helped in the improvement of quality. Sincere thanks to Editor Prof. João M.P.Q. Delgado by the opportunity given to present our research in this book.

References

1. Milman, M.J.: Equipments for Pre-processing of Grain. University Publishing/Federal University of Pelotas (UFPel), Brazil, Pelotas (2002). (In Portuguese)
2. Fioreze, R.: Drying Principles of Organic Products. University Publishing/Federal University of Paraíba (UFPB), João Pessoa (2003). (In Portuguese)
3. Reeds, J.S.: Drying. ASM International Handbook Committee, New York (1991)
4. Brooker, D.B., Bakker-Arkema, F.W., Hall, C.W.: Drying and Storage of Grains and Oilseeds. The AVI Publishing Company, New York (1992)
5. Silva, J.S.: Drying and Storage of Agricultural Products. Learn Easy, Viçosa (2008). (in Portuguese)
6. Peske, S.T., Nedel, J.L., Barros, A.C.S.A.: Rice Production. University Publishing/Federal University of Pelotas (UFPel), Pelotas (1996). (in Portuguese)
7. Kemp, I.C.: Process-system simulation tools. In: Tsotsas E., Mujumdar A.S. (eds.) Modern Drying Technology: Computational Tools at Different Scales, vol. 1, pp. 261–305. Wiley-VHC, Weinheim (2007)
8. Kudra, T., Mujumdar, A.S.: Advanced Drying Technologies. CRC Press, Boca Raton (2007)
9. Mujumdar, A.S, Menon, A.S.: Drying of solids: principles, classification and selection of dryers. In: Mujumdar A.S. (ed.) Handbook of Industrial Drying, vol. 1, pp. 1–40. Marcel Dekker Inc., New York (1995)
10. Dong, R., Lu, Z., Liu, Z., Nishiyama, Y., Cao, W.: Moisture distribution in a rice kernel during tempering drying. J. Food Eng. **91**, 126–132 (2009)
11. Carmo, J.E.F., de Lima, A.G.B., Silva, C.J.: Continuous and intermittent drying (tempering) of oblate spheroidal bodies: modeling and simulation. Int. J. Food Eng. **8**(3), 20 (2012)
12. Carmo, J. E. F.: Transient diffusion phenomena in oblate spheroidal solids: case studies: drying of lentils. Ph.D. Thesis, Process Engineer, Federal University of Campina Grande, Campina Grande (PB), Brazil (2004). (In Portuguese)
13. Chua, K.J., Mujumdar, A.S., Chou, S.K.: Intermittent drying of bio-products—on overview. Bioresource Tchnol. **90**, 285–295 (2003)

14. Váquiro, H.A., Clemente, G., García-Pérez, J.V., Mulet, A., Bon, J.: Enthalpy-driven optimization of intermittent drying of Mangifera indica L. Chem. Eng. Res. Design **87**, 885–898 (2009)
15. Kowalski, S.J., Pawłowski, A.: Energy consumption and quality aspect by intermittent drying. Chem. Eng. Process.: Process Intens. **50**, 384–390 (2011)
16. Baini, R., Langrish, T.A.G.: Choosing an appropriate drying model for intermittent and continuous drying of bananas. J. Food Eng. **79**, 330–343 (2007)
17. Fatouh, M., Metwally, M.N., Helali, A.B., Shedid, M.H.: Herbs drying using a heat pump dryer. Energy Conv. Manag. **47**, 2629–2643 (2006)
18. Aquerreta, J., Iguaz, A., Arroqui, C., Virseda, P.: Effect of high temperature intermittent drying and tempering on rough rice quality. J. Food Eng. **80**, 611–618 (2007)
19. Cihan, A., Ece, M.C.: Liquid diffusion model for intermittent drying of rough rice. J. Food Eng. **49**, 327–331 (2001)
20. Chua, K.J., Chou, S.K., Ho, J.C., Mujumdar, A.S., Hawlader, M.N.A.: Cyclic air temperature drying of guava pieces: Effects on moisture and ascorbic acid contents. Trans. Inst. Chem. Eng. **78**, 72–78 (2000)
21. Dong, R., Lu, Z., Liu, Z., Koide, S., Cao, W.: Effect of drying and tempering on rice fissuring analysed by integrating intra-kernel moisture distribution. J. Food Eng. **97**, 161–167 (2010)
22. Thomkapanich, O., Suvarnakuta, P., Devahastin, S.: Study of intermittent low pressure superheated steam and vacuum drying of heat-sensitive material. Drying Technol. **25**, 205–223 (2007)
23. Holowaty, S.A., Ramallo, L.A., Schmalko, M.E.: Intermittent drying simulation in a deep bed dryer of yerba mate. J. Food Eng. **111**, 110–114 (2012)
24. Putranto, A., Chen, X.D., Xiao, Z., Webley, P.A.: Mathematical modelling of intermittent and convective drying of rice and coffee using the reaction engineering approach (REA). J. Food Eng. **105**, 638–646 (2011)
25. Putranto, A., Xiao, Z., Chen, X.D., Webley, P.A.: Intermittent drying of mango tissues: implementation of the reaction engineering approach. Ind. Eng. Chem. Res. **50**, 1089–1098 (2011)
26. Smith, S.A., Langrish, T.A.G.: Multicomponent solid modeling of continuous and intermittent drying of Pinus radiata sapwood below the fiber saturation point. Drying Technol. **26**(7), 844–854 (2008)
27. Herrithsch, A., Dronfield, J., Nijdam, J.: Intermittent and continuous drying of red-beech timber from the green conditions. In: Proceedings of the 16th International Drying Symposium, Hyderabad, India, 9–12 November (2008)
28. Rémond, R., Perré, P.: High-frequency heating controlled by convective hot air: toward a solution for on-line drying of softwoods. Drying Technol. **26**, 530–536 (2008)
29. Salin, J.G.: A theoretical analysis of timber drying in oscillating climates. Holtzforschung **57**, 427–432 (2003)
30. Kowalski, S.J., Pawłowski, A.: Intermittent drying of initially saturated porous materials. Chem. Eng. Sci. **66**, 1893–1905 (2011)
31. Kowalski, S.J., Pawłowski, A.: Drying of wet materials in intermittent conditions. Drying Technol. **28**, 636–643 (2010)
32. Nishiyama, Y., Cao, W.Y., Li, B.: Grain intermittent drying characteristic analyzed by a simplified model. J. Food Eng. **76**, 272–279 (2006)
33. Jumah, R.Y., Mujumdar, A.S., Raghavan, G.S.V.: A mathematical model for constant and intermittent batch drying of grains in a novel rotating jet spouted bed. Drying Technol. **14**, 765–802 (1996)
34. Garcia, D.C. http://coral.ufsm.br/sementes/textos/secagem.pdf. Accessed 25 Nov 2014
35. Shei, H.J., Chen, Y.L.: Computer simulation on intermittent drying of rough rice. Drying Technol. **20**(3), 615–636 (2002)

36. Sabbah, M.A., Foster, G.H., Haugh, C.G., Peart, R.M.: Effect of tempering after drying on cooling shelled corn. Trans. ASAE **15**(4), 763–765 (1972)
37. Tolaba, M.P., Aguerre, R.J., Suárez, C.: Drying of corn with tempering: simulation and experimental verification. In: IADC—Inter-American Drying Conference, Itu (1997)
38. Steffe, J.F., Singh, R.P., Bakshi, A.S.: Influence of tempering time and cooling on rice milling yields and moisture removal, Trans. ASAE **22**, 1214–1218, 1224 (1979)
39. Walker, L.P., Bakker-Arkema, F.W.: Energy efficiency in concurrent rice drying. Trans. ASAE **24**, 1352–1356 (1981)
40. Fioreze, R.: The intermittent drying agricultural crops with particular reference to energy requirements. Ph.D. Thesis. Granfield Institute of Technology, Silsoe College (1986)
41. Elbert, G., Tolaba, M.P., Suárez, C.: Effects of drying and tempering on head parboiled rice yield. In: IADC—Inter-American Drying Conference, Itu, Brazil, 15–18 July (1997)
42. Steffe, J.F., Singh, R.P.: Theoretical and practical aspects of rough rice tempering. Trans. ASAE **23**(3), 775–782 (1980)
43. Mabrouk, S.B., Benali, E., Oueslati, H.: Experimental study and numerical modelling of drying characteristics of apple slices. Food Bioprod. Process. **90**, 719–728 (2012)
44. Silva, V., Figueiredo, A.R., Costa, J.J., Guiné, R.P.F.: Experimental and mathematical study of the discontinuous drying kinetics of pears. J. Food Eng. **134**, 30–36 (2014)
45. Poomsa-ad, N., Soponronnarit, S., Prachayawarakorn, S., Terdyothin, A.: Effect of tempering on subsequent drying of paddy using fluidisation technique. Drying Technol. **20**(1), 195–210 (2002)
46. Jaisut, D., Prachayawarakorn, S., Varanyanond, W., Tungkrakul, P., Soponronnarit, S.: Effects of drying temperature and tempering time on starch digestibility of brown fragrant rice. J. Food Eng. **86**(2), 251–258 (2008)
47. Law, C.L., Mujumdar, A.S.: Dehydration of Fruits and Vegetables, In Guide to Industrial Drying. IDS, Hyderabad (2008)
48. Chin, S.K., Law, C.L.: Product quality and drying characteristics of intermittent heat pump drying of Ganoderma tsugae Murrill. Drying Technol. **28**, 1457–1465 (2010)
49. Chua, K.J., Chou, S.K., Mujumdar, A.S., Hawlader, M.N.A., Ho, J.C.: Heat pump drying of banana, guava, and potato pieces: effect of cyclical variations of air temperature on convective drying kinetics and color change. Drying Technol. **18**(5), 907–936 (2000)
50. Stratton, J.A., Morse, P.M., Chu, L.J., Little, J.D.C., Huntner, R.A.: Elliptic Cylinder and Spheroidal Wave Functions. The Technical Press of M. I. T. and Wiley, New York (1941)
51. Flammer, C.: Spheroidal Wave Functions. Stanford University Press, Stanford (1957)
52. Abramowitz, M., Stegun, I.A.: Handbook of Mathematical Functions, pp. 752–772. Dover Publications Inc., New York (1972)
53. Whitaker, S.: Heat and Mass Transfer in Granular Porous Media, In Advances in Drying. Hemisphere Publishing Corporation, New York (1980)
54. Patankar, S.V.: Numerical Heat Transfer and Fluid Flow. Hemisphere Publishing Coorporation, New York (1980)
55. Maliska, C.R.: Computational Heat Transfer and Fluid Mechanics. LTC, Rio de Janeiro (1995). (In Portuguese)
56. Versteeg, H.K., Malalasekera, W.: An Introduction to Computational Fluid Dynamics: The Finite Volume Method. Prentice Hall, London (1995)
57. Lima, A.G.B., Nebra, S.A., Altemani, C.A.C.: Simulation of the drying kinetics of the silkworm cocoon considering diffusive mechanism in elliptical coordinate. In: Inter-American Drying Conference (IADC 1997), Itu, Brazil (1997)
58. Lima, A.G.B.: Diffusion phenomenon in prolate spheroidal solids. Case studies: drying of banana. Ph.D. Thesis, Mechanical Engineering Faculty, State University of Campinas, Campinas (SP) Brazil (1999). (in Portuguese)
59. Lima, A.G.B., Nebra, S.A.: Theoretical analysis of the diffusion process inside prolate spheroidal solids. Drying Technol. **18**(1–2), 21–48 (2000)

60. Carmo, J.E.F., Lima, A.G.B.: Modelling and simulation of mass transfer inside the oblate spheroidal solids. In: Ducere S.A. (ed.) 2nd Interamerican Drying Conference, Boca del Rio, Vera Cruz (IADC2001), vol. 1, pp. 173–183 (2001)
61. Carmo, J.E.F., Lima, A.G.B.: Drying of lentil including shrinkage: a numerical simulation. Drying Technol. **23**, 1977–1992 (2005)
62. Carmo, J.E.F., Lima, A.G.B.: Mass transfer inside oblate spheroidal solids: modelling and simulation. Braz. J. Chem. Eng. **25**(1), 1–8 (2008)

Clay Products Convective Drying: Foundations, Modeling and Applications

A.G. Barbosa de Lima, J. Barbosa da Silva, G.S. Almeida, J.J.S. Nascimento, F.V.S. Tavares and V.S. Silva

Abstract This chapter briefly focuses on the theory and applications of drying process with particular reference to wet clay product. Herein, a modeling based on the heat and liquid diffusion theories including dimensions variations and hygro-thermal-elastic stress analysis, and the mathematical formalism to obtain the numerical solution of the governing equations using finite-volume method are presented. The model considers constant thermo-physical properties and convective boundary conditions at the surface of the solid. Applications have been done to ceramic hollow brick. Predicted results of the average moisture content, surface temperature, and moisture content, temperature and stress distributions within the porous solids are shown and analyzed, and for some drying situations they are compared with experimental drying data of the average moisture content and surface temperature of the brick along the continuous drying process.

A.G. Barbosa de Lima (✉) · J.B. da Silva · F.V.S. Tavares · V.S. Silva
Department of Mechanical Engineering, Federal University of Campina Grande, Av. Aprígio Veloso, 882, Bodocongó, 58429-900 Campina Grande, PB, Brazil
e-mail: antonio.gilson@ufcg.edu.br

J.B. da Silva
e-mail: barbosa.joselito@gmail.com

F.V.S. Tavares
e-mail: valdeiza.tavares@hotmail.com

V.S. Silva
e-mail: veralucia.silva84@yahoo.com.br

G.S. Almeida
Federal Institute of Education, Science and Technology of Roraima, IFRR, Boa Vista-RR, Brazil
e-mail: genival.almeida@ifrr.edu.br

J.J.S. Nascimento
Department of Materials Engineering, Federal University of Campina Grande, Av. Aprígio Veloso, 882, Bodocongó, Campina Grande, PB 58429-900, Brazil
e-mail: jefferson@dema.ufcg.edu.br

J.M.P.Q. Delgado and A.G. Barbosa de Lima (eds.),
Drying and Energy Technologies, Advanced Structured Materials 63,
DOI 10.1007/978-3-319-19767-8_3

Keywords Drying · Numerical · Experimental · Hollow brick

1 Introduction

Ceramic processing has traditionally been discussed in term of the material formulation and industrial arts used in the production of commercial products that are very different in size, shape, detail, complexity, material composition, structure, and cost. Structural clay products include e.g. building bricks and tiles (traditional ceramics) where in which structural intensity is very important [1, 2]. One important concern in the application of ceramic materials is the fabrication technique. Clay minerals, when mixed with water, become highly plastic (hydroplasticity) and it can be molded at different shapes easily without cracking. These physical and chemical characteristics maintain the body shape during handling, drying and firing. The water added to ceramic clay cover the surface of the clay particle (forming a thin film around the clay particles), fills small capillaries and porous, and causes a separation of particles inside it [3]. Then, the particles are free to moving over another resulting in plasticity of the water-clay system.

After plastic forming and casting, products must be dried, in order, to removal of the moisture prior for further processing and/or firing. If this moisture is not entirely removed, the extreme temperature in the kiln will force out this water during firing, causing cracking and until explosion of the product. In the case of argils, to promote an understanding of the drying mechanism, a microscopic investigation in the state of the moisture inside the product it is requested [4–6].

In this sense, some authors they affirm that during clay drying, the dominant mechanism of moisture migration is the liquid transport [7–12], while another they consider that exist liquid and vapor transport inside the solid [13–15]. On the basis of their hypothesis several studies on clay drying have been reported in the literature [16–26].

Drying is a simultaneous of heat and mass transfer including dimension variations. As drying progresses and water is removed, the interparticle separation decreases, a condition termed shrinkage [1, 2]. During the drying of ceramics products, heat is transported to the liquid inside the solid, and simultaneously moisture is transported to out of solid, generating dilation by heating and contraction by moisture removal.

Drying of ceramics is accomplished by evaporation of water. If the rate of liquid water evaporation at the surface of the clay body is greater than the rate of liquid water diffusion inside the solid, the surface will dry more quickly than the interior by causing stresses and probable defects or perhaps revealing them.

These mechanisms of heat and mass transfer are more intense in ceramic materials with high initial moisture content, mainly in products of fine granulation and gives rise to hygro-thermal stresses on the product. The magnitude of these stresses will depend on the shape, thermo-physical and elastic properties, and contraction and expansion coefficients of the material.

Different factors like area/volume relationships, initial water content, granulometry (clay particle size) and shape of the formed body, and air-drying conditions (temperature, relative humidity and flow rate) affect shrinkage phenomenon. Then, it is critical to control the rate of water removal during clay materials drying [4–6]. These defects range from visible defects (such as cracks) to a reduction in physical properties in fired clay products in areas such as strength, elastic modulus, color and appearance (discoloration due to summing) [3]. Beside, drying represents a process cost to the manufacturer in terms of handling and in terms of energy costs to accomplish drying. Then, deformation and failure of clay drying are great problems in a ceramic production. Therefore the stress control during drying process is relevant for the quality of the product after drying and for energy saving.

2 Clay Drying

2.1 Foundations

Brick is a commercial ceramic product made of clay minerals and has been used in worldwide for centuries in house wall and building constructions. The ceramic product, during the manufacturing process, goes through the steps of product shaping (molding), drying and firing [1–3]. The drying is the phase of the process that precedes firing and needs an appreciable amount of thermal energy to evaporate the water which was added during the molding process [1, 6]. The purpose of this step is to reduce slowly and uniformly, the moisture of the products from 20–25 %, after extrusion or pressing, to 3–10 % at the end of drying. Drying causes shrinkage of product which can vary, in general, from 4 to 10 %. In case of the ceramic products, where red clay is used, the drying stage has a great importance. In this step most of the water contained in the product is eliminated. Consequently higher shrinkage and dilation of material results and the product often gets rejected due to its low quality. The published [3, 6] literature shows that if a special care is taken during the drying step energy cost can be reduced and a better quality product can be obtained.

When the ceramic product drying operation is natural, the pieces are stacked in covered sheds, arranged on shelves (fixed or mobile) or simply piled on the floor. The duration of drying depends on the material characteristics and state conditions of atmospheric air (temperature and relative humidity) and of the ventilation of the place, needing periods of up to 6 weeks. However, to reduce drying time, artificial drying is carried out in drying chambers or ovens, using, as a rule, the residual heat of the furnaces, while these are being cooled. The most common types of artificial dryers are static, continuous or semi continuous. The period of artificial drying depends on the characteristics of the raw material, the shape of the pieces to be dried and the type of dryer and air drying conditions. It has an average variation of 12–40 h. Artificial drying is carried out at temperature 80–110 °C, and precise information about the relative humidity are practically non-existent in the literature, though it can be said that this is very high [5, 27].

The choice of a particular type of dryer for a particular drying operation is a complex task by virtue of the existence of many factors that affect the selection. In particular, factors to be considered during the drying process includes the drying technique, heating method and the mass and energy exchanges mechanisms between the product and the drying air. Currently certain preference is given to tunnel and batch (type oven) dryers for drying of ceramic materials, specifically red ceramic one. Within this context, for the purpose of searching new technologies for the sector, improvement of the product quality and rationalization of energy consumption, several studies have been conducted [12, 28–32].

Because of the importance and cost of the drying procedure, mainly in industrial scale mathematical models to describe the drying process of ceramic material have been reported by many researchers [6, 7, 9, 10, 12–15, 24, 33–41].

Depending on the thickness of the material studied, these models can be classified into: drying in a thin layer (particle models) and drying in a thick layer (dryer level models). The drying of a single particle individually or even of a layer of material of small thickness, does not modify the drying conditions significantly as the air past over the wet porous solid. The practical importance of thin layer drying is very limited, because the materials are generally dried in thick layers: stationary or moving. When the material is placed inside of a dryer, forming a thick layer, the thermodynamic properties of the drying air are modified markedly as the air flows through it. In virtue of this, the models of drying in thick layer, having more complex equations that take into account the heat and mass transfer between the product and the surrounding air, are more complete than the thin layer drying model.

Despite of the importance on the dryer model in the chapter special attention should be done to a robust particle model with particular reference to industrial hollow brick drying.

2.2 Mathematical Modeling: Application to Hollow Brick

Drying characteristics of the particular materials being dried and simulation models are needed in the design, construction and operation of drying systems. Several empirical, semi-empirical, and theoretical models have been proposed to predict drying process of different materials. For convective drying process, theoretical models take into account both internal and external resistance while the semi-empirical and empirical models (lumped models) take into account only external resistance to heat and moisture transfer between porous solid and air surrounding it (internal resistance is negligible). At present, there are very few models that represent the batch drying of clay products specially brick. As shrinkage in clay brick is an observable phenomenon during drying process and may have significant effect on mass diffusivity, it must be taken into account in order to obtain reliable predictions of performance. Despite of the importance, because the simplifications, the lumped models cannot give clear and accurate of the fundamentals of the drying process, but they can used to complete mathematical models at the level of the drying equipment.

Mass transfer during falling rate period is caused by liquid diffusion or capillary flow. The former is commonly used to describe drying behavior in the falling rate period of clay products. In the liquid diffusion theory [42–44], the rate of diffusion is governed by moisture concentration gradient as the driving force. Fick's law of diffusion is widely used to model the drying behavior for this period.

Heat flux inside the wet clay porous product during drying process can be modeled by considering heat conduction as unique mechanism of heat transfer. In this case we can use the Fourier´s law to model this heating process.

To describe the liquid and heat transfer inside the clay brick, the following assumptions were used:

- The thermo-physical properties are constant, during the whole drying process;
- Source term of energy and mass is neglected;
- The solid is homogeneous and isotropic;
- The moisture content and temperature distributions are uniform at the beginning of the process;
- The phenomenon happens under convective condition at the surface of the body.
- The shrinkage of the solid is polynomial function of the moisture content;
- The solid is composed of water in liquid phase and solid material.
- The diffusion phenomenon occurs under falling rate.

Figure 1 illustrates a solid parallelepiped of dimensions $2R_1 \times 2R_2 \times 2R_3$. For this case, the general partial differential equations that describe the unsteady diffusion phenomenon in a three-dimensional situation it is in the way as follows

2.2.1 Mass Transfer Model

For mass transfer in the absence of mass source, the Fick´s law was used as follows:

$$\frac{\partial M}{\partial t} = \nabla \cdot (D \nabla M) \tag{1}$$

where M (moisture content, dry basis), D represents mass diffusion coefficient and t is the time.

Due to the symmetry in the solid, in the planes (x = 0, y, z), (x, y = 0, z), (x, y, z = 0) it was considered 1/8 of the total volume of the solid. Therefore, the initial, symmetry and boundary conditions are as follows:

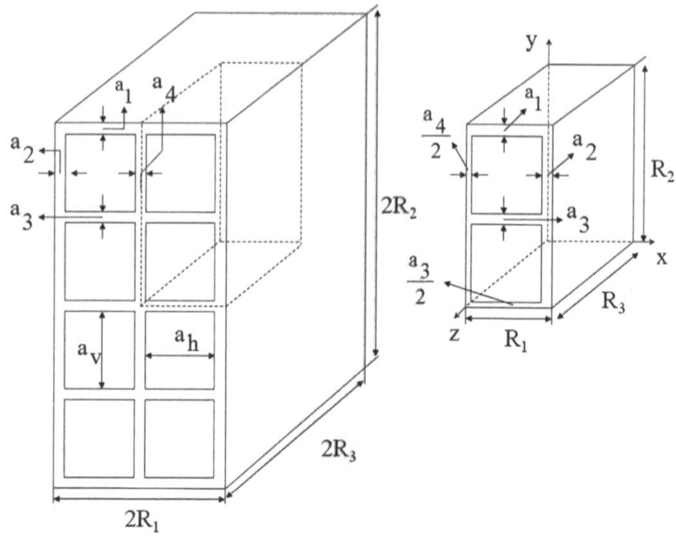

Fig. 1 Geometrical parameters of a specific hollow brick

- Initial condition:

$$M(x, y, z, t = 0) = M_o \quad (2)$$

- Symmetry condition:

$$\frac{\partial M(x = 0, y, z, t)}{\partial x} = \frac{\partial M(x, y = 0, z, t)}{\partial y} = \frac{\partial M(x, y, z = 0, t)}{\partial z} = 0, \ t > 0 \quad (3)$$

- Boundary condition:

$$-D\frac{\partial M(x = R_1, y, z, t)}{\partial x} = h_m[M(x = R_1, y, z, t) - M_e] \quad (4)$$

$$-D\frac{\partial M(x, y = R_2, z, t)}{\partial y} = h_m[M(x, y = R_2, z, t) - M_e] \quad (5)$$

$$-D\frac{\partial M(x, y, z = R_3, t)}{\partial z} = h_m[M(x, y, z = R_3, t) - M_e] \quad (6)$$

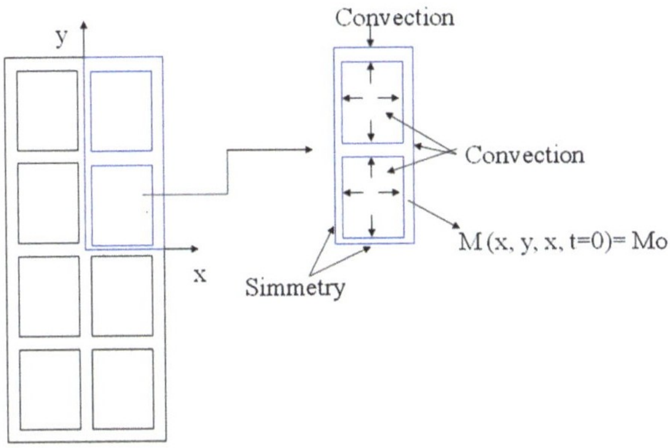

Fig. 2 Scheme showing the boundary conditions in the hollow brick

where h_m correspond to convective mass transfer coefficient and M_e represents equilibrium moisture content of the porous solid with the drying air.

For convenience, Fig. 2 illustrates better the initial, symmetry and boundary conditions.

2.2.2 Heat Transfer Model

For heat transfer, in the absence of energy source term, the Fourier´s law was used as follows:

$$\frac{\partial \theta}{\partial t} = \nabla \cdot (\alpha \nabla \theta) \tag{7}$$

where θ is the solid temperature and $\alpha = k/(\rho c_p)$ represent thermal diffusivity, being k the thermal conductivity, c_p the specific heat and ρ is the solid density.

The initial, symmetry and boundary conditions are as follows:

- Initial condition:

$$\theta(x, y, z, t = 0) = \theta_o \tag{8}$$

- Symmetry condition:

$$\frac{\partial \theta(x = 0, y, z, t)}{\partial x} = \frac{\partial \theta(x, y = 0, z, t)}{\partial y} = \frac{\partial \theta(x, y, z = 0, t)}{\partial z} = 0, \ t > 0 \tag{9}$$

- Boundary condition:

$$-k\frac{\partial\theta(x = R_1, y, z, t)}{\partial x} = h_c[\theta(x = R_1, y, z, t) - \theta_e] \quad (10)$$

$$-k\frac{\partial\theta(x, y = R_2, z, t)}{\partial y} = h_c[\theta(x, y = R_2, z, t) - \theta_e] \quad (11)$$

$$-k\frac{\partial\theta(x, y, z = R_3, t)}{\partial z} = h_c[\theta(x, y, z = R_3, t) - \theta_e] \quad (12)$$

where h_c correspond to convective heat transfer coefficient and θ_e represents equilibrium temperature between the solid and drying air.

The average value of moisture content or temperature was obtained as follows:

$$\bar{\Phi} = \int_V \Phi dV \quad (13)$$

where $\Phi = \theta$ (temperature) or $\Phi = M$ (moisture content, dry basis) and V is the volume of the solid.

2.2.3 Volume Variations Model

During drying process, the wet porous solid surfer expansion (dilation) by heating and contraction (shrinkage) by loss of water. Dilation phenomenon has few effect on the dimension variation of the solid because the volumetric dilation coefficient is smallest than the shrinkage coefficient. Thus, following equation was utilized to obtain the changes in volume (shrinkage) in each time during the drying process:

$$V_t = V_o\left(b_1\overline{M}^3 + b_2\overline{M}^2 + b_3\overline{M} + b_4\right) \quad (14)$$

where V_o is initial volume of the brick and b_1, b_2, b_3 and b_4 are parameters to be determined by fitting to experimental data.

The volume of the hollow brick is given by:

$$(V)_t = 8[R_1R_2R_3 - 2a_va_hR_3] \quad (15)$$

where $a_v = [2R_2 - (2a_1 + 3a_3)]/4$ and $a_h = [2R_1 - (2a_2 + a_4)]/2$.

The surface area of heat and mass transfer of the solid during the drying process was obtained by:

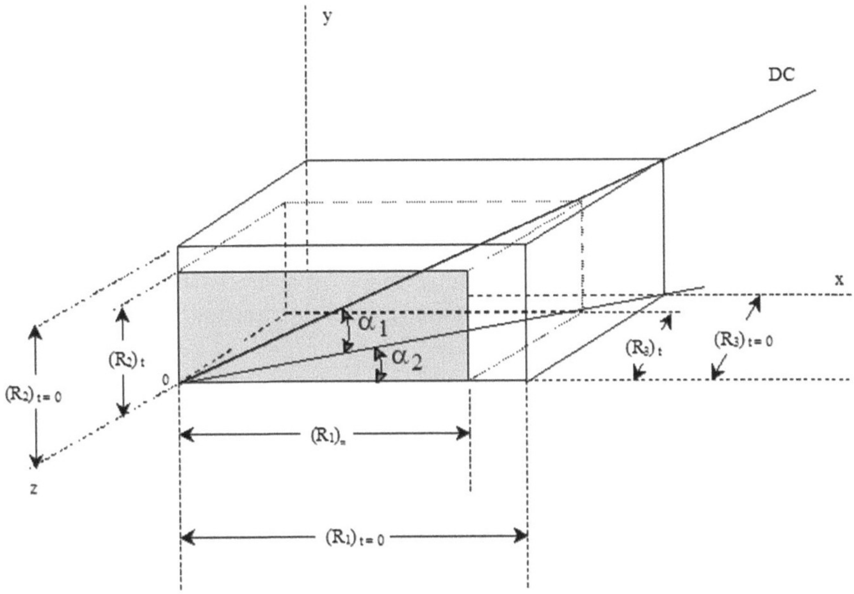

Fig. 3 Shrinkage of a solid parallelepiped during the drying process

$$A_s = 8[R_1R_2 + R_2R_3 + R_1R_3 - 2a_va_h + 4a_vR_3 + 4a_hR_3] \tag{16}$$

For the calculation of the new volume and surface area of the brick during drying process considers the Fig. 3. Based on this figure was assumed an isotropic shrinkage. Thus, is possible to demonstrate that:

$$\tan\alpha_2 = \frac{(R_3)_t}{(R_1)_t} = \frac{(R_3)_{t=0}}{(R_1)_{t-0}} = \hat{k}1 \tag{17}$$

$$\tan\alpha_1 = \frac{(R_2)_t}{\sqrt{(R_1)_t^2 + (R_3)_t^2}} = \frac{(R_2)_{t=0}}{\sqrt{(R_1)_{t=0}^2 + (R_3)_{t=0}^2}} = \hat{k}2 \tag{18}$$

where $\hat{k}1$ and $\hat{k}2$ are constant. So by combining Eqs. (14), (17) and (18) the new dimension of the porous body can be determined in each instant of the drying process.

2.2.4 Stress Model

The governing equation in a stress field derived from force balance for static equilibrium can be expressed as follows:

$$\nabla\sigma + \vec{F} = 0 \tag{19}$$

where σ is a stress tensor and $\vec{F}$ is a body force vector.

At the surface, the boundary conditions can be described as follows:

$$u = u_a \text{ and } \sigma \cdot \vec{n} = t \tag{20}$$

where u_a is a prescribed displacement and t is a traction on the boundary.

The basic assumption of a constitutive model in a stress field is that a strain tensor represents the sum of the strains caused by traction, temperature and moisture as follows:

$$e = \frac{1}{E}(\sigma - \sigma_o) + \gamma\Delta\theta + \beta\Delta M \tag{21}$$

where e is a total mechanical strain tensor, σ is a total stress tensor, σ_o is an initial stress tensor, E is an elastic coefficient, γ is a thermal expansion coefficient, $\Delta\theta = \theta - \theta_o$ is a temperature variation, β is a moisture expansion coefficient (effective shrinkage coefficient) and $\Delta M = M - M_o$ is a moisture content variation. Stresses caused by thermal expansion (heating) are opposed those caused by drying shrinkage (moisture removal).

In a hygro-thermal stress problem, the thermal and moisture effects may be treated as that of an initial stress. The constitutive equation based in the Eq. (21) can be defined as follows [45–47].

$$\sigma = E(e - \gamma\Delta\theta - 0\Delta M) + \sigma_o = Ee + \sigma_o' \tag{22}$$

According to elasticity theory [48], there are three principal stresses formed at any location in a strained body (Fig. 4). These stresses are oriented orthogonally to each other, and unidirectional stresses (tensile or compressive stresses) are locally maximum along these orientations [49]. Based on this theory, the following assumptions were adopted:

- No initial stress σ_o are applied to the brick;
- Mechanical strain are neglected ($\varepsilon_x = \varepsilon_y = \varepsilon_z = e = 0$);
- Internal stresses (σ_x, σ_y and σ_z) are due to hygro and thermal strains due to dilation and shrinkage;
- Hygro and thermal strain causes expansion and contraction and the change volume element. No shear exists. This way the shape of the element is unchanged. The body is unrestrained (free to expand and constrains).
- Because no shear stress occurs, the stresses (σ_x, σ_y and σ_z) are the principal stresses (maximum normal stress in each location in the brick).

According to these considerations, the total stress inside the brick is the sum of hygro and thermal stresses. So, by using Eq. (22), the normal stress in the x, y and z directions can be calculated using the following equations:

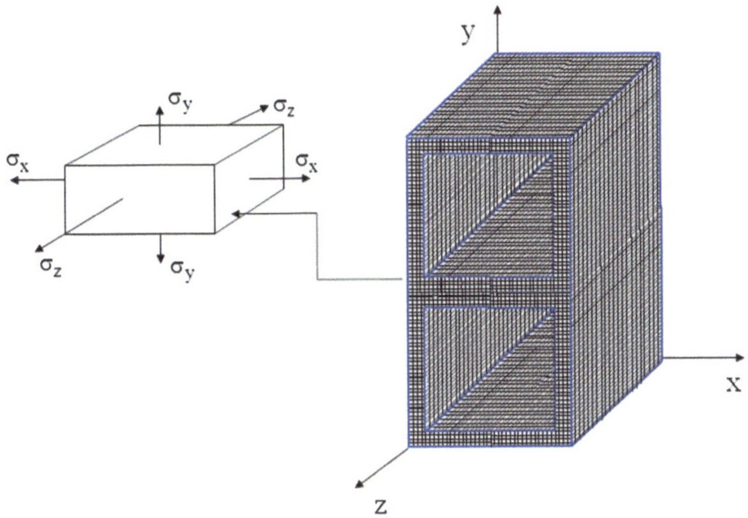

Fig. 4 Principal stresses σ_x, σ_y, and σ_z applied in a location inside the brick

$$(-\alpha_{\theta x}\Delta\theta - \alpha_{Mx}\Delta M)E = [\sigma_x - \nu(\sigma_y + \sigma_z)] \tag{23}$$

$$(-\alpha_{\theta y}\Delta\theta - \alpha_{My}\Delta M)E = [\sigma_y - \nu(\sigma_x + \sigma_z)] \tag{24}$$

$$(-\alpha_{\theta z}\Delta\theta - \alpha_{Mz}\Delta M)E = [\sigma_z - \nu(\sigma_x + \sigma_y)] \tag{25}$$

where ν is the Poisson coefficient, $\alpha_{\theta x} = \alpha_{\theta y} = \alpha_{\theta z} = \alpha_\theta$ are the linear thermal expansion coefficients and $\alpha_{Mx} = \alpha_{My} = \alpha_{Mz} = \alpha_M$ are the linear hydrical expansion coefficients. Solving the linear equation system (Eqs. 23–25) σ_x, σ_y and σ_z can be determined as follows:

$$\sigma_x = \sigma_y = \frac{[(-\alpha_\theta\Delta\theta - \alpha_M\Delta M)E + \nu\sigma_z]}{1-\nu} \tag{26}$$

$$\sigma_z = \frac{2E(-\alpha_\theta\Delta\theta - \alpha_M\Delta M)}{(2\nu - 1)} \tag{27}$$

2.3 *Numerical Solution*

Literature reports different exact (separation of variables, Laplace Transform, Galerkin-based integral method, etc.) and numerical (finite difference, finite element, boundary element, finite volume, etc.) methods to solve a partial differential equation. In this chapter was used the finite-volume method to discretize the basic

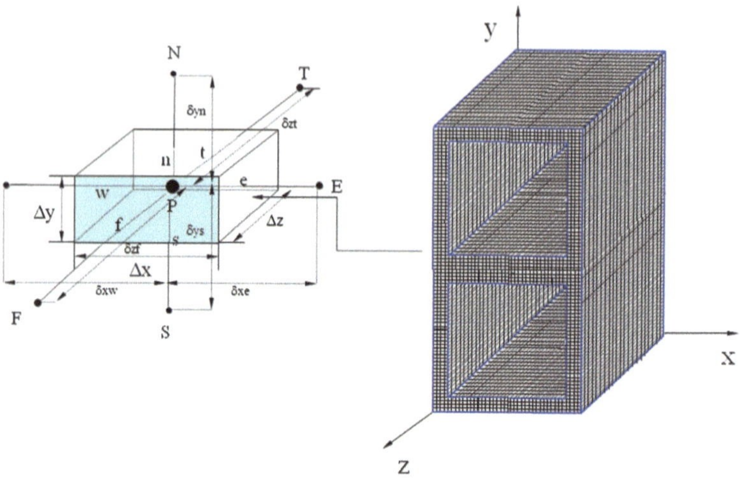

Fig. 5 Control-volume and grid used in this work

equations. Now a three-dimensional grid is used to subdivide the domain. The cell contain node P and six neighboring nodes identified as west, east, south, north, bottom and top nodes. Figure 5 represents the differential volume of the physical domain (Fig. 1), where the nodal points (W, E, N, S, F, T) and geometrical parameters of the control volume are presented.

Assuming fully implicit scheme, where all terms are estimated in t + Δt, the Eqs. (1) and (7) were integrated in the control volume of the Fig. 5, that correspond to the internal points of the domain, and also in the time. As results of this mathematical procedure the following discretization equation [50–52] in the short form was obtained:

$$A_P\Phi_P = A_E\Phi_E + A_W\Phi_W + A_N\Phi_N + A_S\Phi_S + A_T\Phi_T + A_F\Phi_F + B \tag{28}$$

The implicit method is recommended for general purpose transient calculations because of its robustness and unconditional stability. Discretized equation must be set up at each of the nodal points in order to solve a problem. Starting with initial field for the potential Φ (mass or temperature) applied to all nodes the set of equations is solved iteratively until a converged solution is obtained. Here the set of equations are solved iteratively using the Gauss-Seidel method. The following convergence criterion was used.

$$\left|\Phi^{n+1} - \Phi^{n}\right| \leq 10^{-8} \tag{29}$$

where n represents the n-th iteration in each time. Details about the numerical procedure can be found in [9–11, 23, 27].

2.4 Experimental Study of Hollow Brick

For the drying experiment, industrial hollow brick (clay) was dried in an oven under controlled conditions of air (velocity, temperature and relative humidity). The geometrical parameters and initial conditions of the material are show in the Table 1. The sample was weighed and measured dimensions and surface temperature at 10 min intervals during the drying process. Figure 6 illustrates a scheme of the clay brick used in this work indicating the point where surface temperature was measured. The continuous drying experiments ended when the mass reached constant weight. In order, to obtain the balanced moisture content, the sample was kept under the same temperature for 48 h inside the oven. The test was performed under atmospheric pressure. Details about the experimental procedure can be found in [23, 25].

2.5 Results Analysis

2.5.1 Experimental

The collected data of the chemical analyses of clay used for brick molding are show in the Table 2. There data suggest that the clay of the Brazilian region is basically rich in aluminum (Al_2O_3), Silica (SiO_2), and iron oxide (Fe_2O_3).

In order to analyze the effects of drying air conditions on the moisture removal and heating of ceramic bricks, experimental data of the dimensionless average moisture content, dimensionless surface temperature (vertex) and volume variations of the brick obtained during the drying were plotted for two drying conditions as seen in the Figs. 7, 8, 9, 10, 11 and 12.

By comparing the data, one could see that the drying air temperature has influenced strongly the drying rate as expected. When the temperature increases, so does the drying rate increase too. However, this may give rise to high temperatures and thermal and hydric gradients into the solid, which would induce thermal, hydrical and mechanical stresses (Figs. 10, 11, 12 and 13). The highest moisture content and temperature gradients occur at the surface of the solid mainly close to the vertexes of the solid, rendering theses regions more vulnerable to cracks and fissures. It can be seen that the stresses have produced serious cracks, fissures and deformations in the brick contributing, as a result, towards a decrease in the quality of the solid at the end of drying process. The mechanical behavior of clay is generally described as elastic viscosity and plasticity depending on the moisture content. Further, the drying rate can also be affected by the shape of the brick and air relative humidity. For higher area/volume relationships and lower air relative humidity we have, as expected, higher drying and shrinkage velocities. It was observed that the y-axis shrink more than x-axis due to the fact of $R_2 > R_1$. The shrinkage is proportional the dimensions of the solid, so, different deformations and

Table 1 Experimental parameters of the clay brick and drying-air

Air			Holed clay brick											
RH (%)	T (°C)	v (m/s)	$2R_1$ (mm)	$2R_2$ (mm)	$2R_3$ (mm)	a_1 (mm)	a_2 (mm)	a_3 (mm)	a_4 (mm)	M_o (d.b.)[a]	M_e (d.b.)	θ_0 (°C)	θ_f (°C)	t (h)
4.6	80.0	1.0	92.76	197.0	201.0	8.16	7.20	7.84	6.66	0.152	0.00039	21.4	69.2	15.0
1.8	100.0	1.0	92.80	198.0	202.0	11.70	9.41	8.74	8.00	0.169	0.00038	26.1	93.2	12.3

[a]d.b. (dry basis)—kg water/kg dry solid

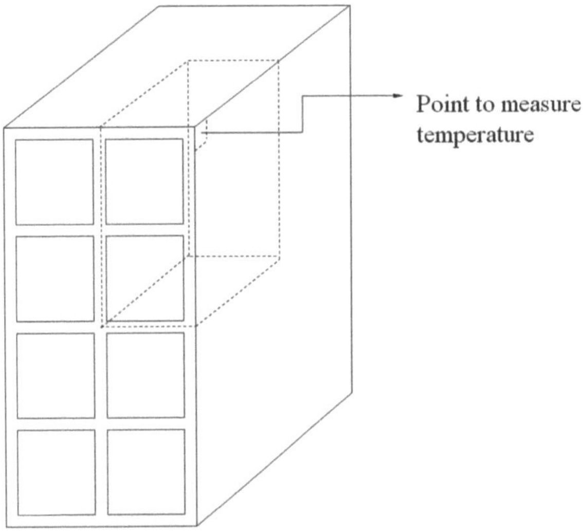

Fig. 6 The geometry of industrial brick

shrinkage velocity occurs in the x, y and z directions. According to Figs. 9 and 10 can be verified that shrinkage is not linear and become more intense as increasing air temperature. Volume variations closed to 13 and 22 %, were obtained close to for temperatures 80 and 100 °C, respectively.

2.5.2 Simulations

As purpose to evaluate the robustness of the mathematical modeling presented before, this one was used to predict hollow brick drying process. Simulations were performed using a grid with 20 × 20 × 20 nodal points and $\Delta t = 1$ s. These conditions were obtained by successive grid and time step refinements. The following data were used: $D = 8.0 \times 10^{-10}$ m²/s; $\alpha = k/\rho c_p = 1.0 \times 10^{-7}$ m²/s; $h_m = 1.0 \times 10^{30}$ m/s; $h_c = 2.41$ W/m² K; $\alpha_M = \beta/3 = 3.3 \times 10^{-2}$; $\alpha_\theta = \gamma/3 = 6.0 \times 10^{-6}$ °C^{-1}; $\theta_e = 80$ °C; $\theta_o = 25$ °C; $\nu = 0.35$; E = 70.0 MPa.

Herein, one comparison between numerical and experimental data of the dimensionless average moisture content and dimensionless surface temperature of the hollow brick obtained during the drying (T = 80 °C) are illustrate in Fig. 14. The mass transport coefficients (D and h_m) were estimated by using the least square error technique. The error and variance obtained for moisture content and temperature were $ERMQ_M = 0.259069$ (kg water/kg dry solid)² and $\overline{S}_M^2 = 0.181$ % kg water/kg dry solid and $ERMQ_T = 0.259069$ (°C/°C)² and $\overline{S}_T^2 = 0.181$ % °C/°C, respectively. The small error and variance indicates that the model presents good agreement with the experimental data. Some discrepancies appear at lowest moisture content and highest surface temperatures due to the fact that for longer drying

Table 2 Chemical composition of the clay (%)

SiO_2	Al_2O_3	Fe_2O_3	K_2O	MgO	CaO	TiO_2	SO_3	MnO	P_2O_5	ZrO_2	SrO	Rb_2O
57.054	24.25	9.257	2.879	2.789	2.049	1.259	0.149	0.140	0.071	0.048	0.034	0.021

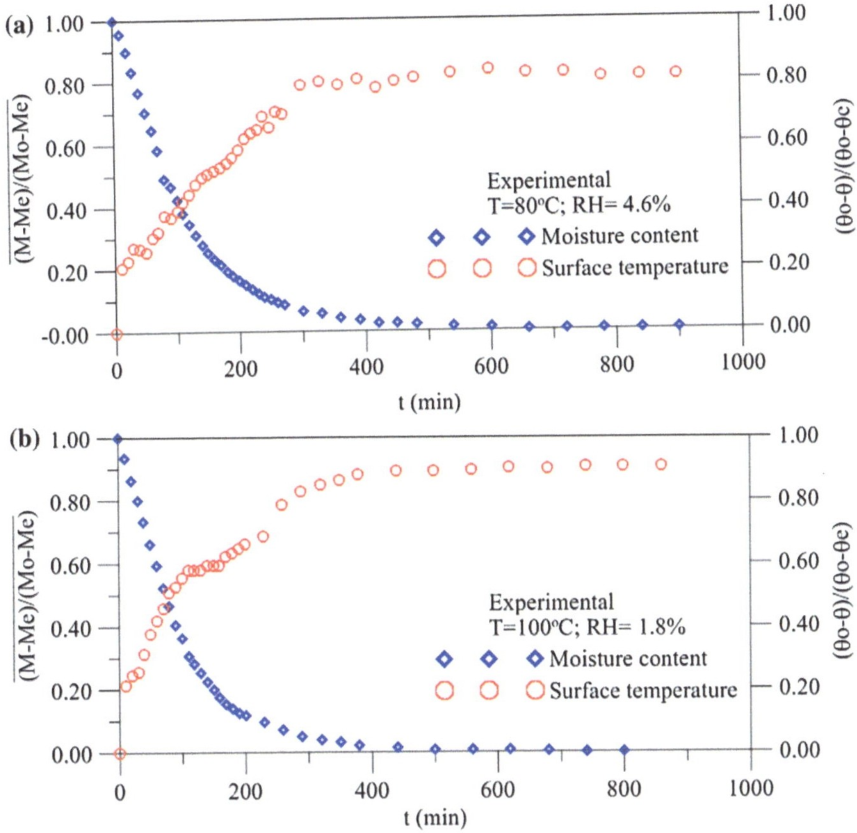

Fig. 7 Drying and heating kinetics of the industrial brick dried to the **a** T = 80 °C and **b** T = 100 °C

times the assumptions of constant thermal properties and polynomial shrinkage are probably not a good choice for this physical problem.

The drying parameters distributions inside the solid play important role to study the evolution of stresses, developed into the wet porous body during process. Then, moisture content, temperature and normal stresses distributions inside the solid obtained during drying ($z \cong R_3/2$) for elapsed times t = 300, 2000 and 8000 s are illustrate in Figs. 15, 16, 17. It is verified that highest moisture content and temperature gradients occurs at the surface of the solid mainly in the vertexes. So, these regions are more susceptible to cracks, fissures and deformations that reduce quality of the bricks.

It is verified that the drying induced stresses increased nonlinearly with the ΔM and Δθ. The maximum stresses, therefore, may occur in the early stage of drying as low initial moisture content of the brick and when higher air temperature and lower relative humidity are used. In the beginning of the drying, because large

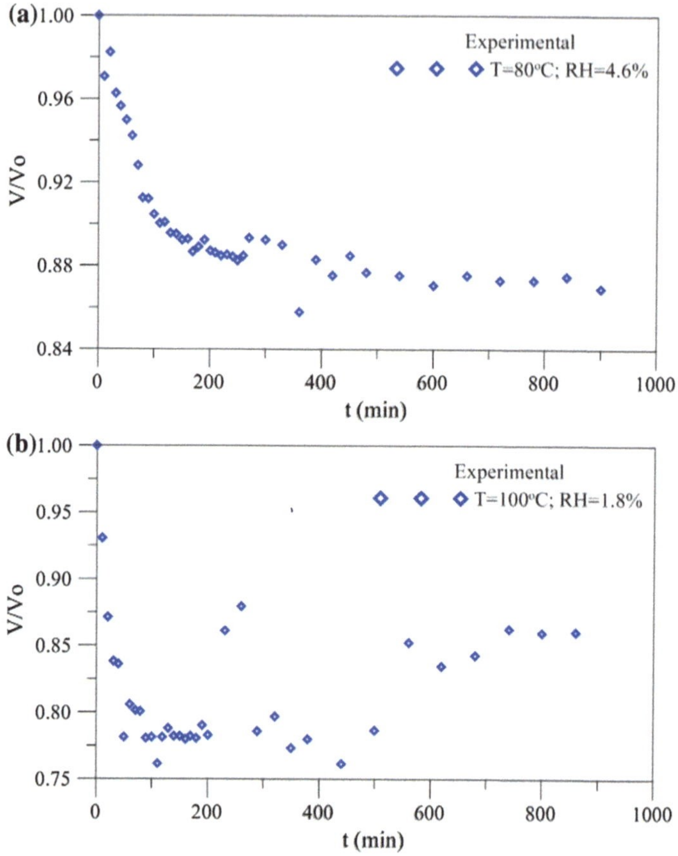

Fig. 8 Dimensionless volume of the clay brick as a function of drying time. **a** T = 80 °C and **b** T = 100 °C

temperature variations, the thermal stress are more than hydrical stress. For large elapsed time, both the hygro and thermal stresses vanish to equilibrium, and total stresses are of tensile in the center and surface. The maximum principal stress has occurred inside the hollow brick by combination of the heating and moisture migration effects.

According to literature [17], if the drying of molded clay is made to proceed too quickly, the result is that the surface dries up immediately while maintaining a wet state inside of product, the drying rate falls markedly and the product may develop cracks. Then, drying must be carried out very slowly under air controlled conditions (moderate temperature and high relative humidity) in order, to prevent undesirable cracks and deformations during the process and thus, obtaining product with good quality post-drying.

Research about strain and stress in clay during drying process has been reported in the literature [4, 14, 17, 36, 53, 54]. For common brick [4, 20, 53, 54], the

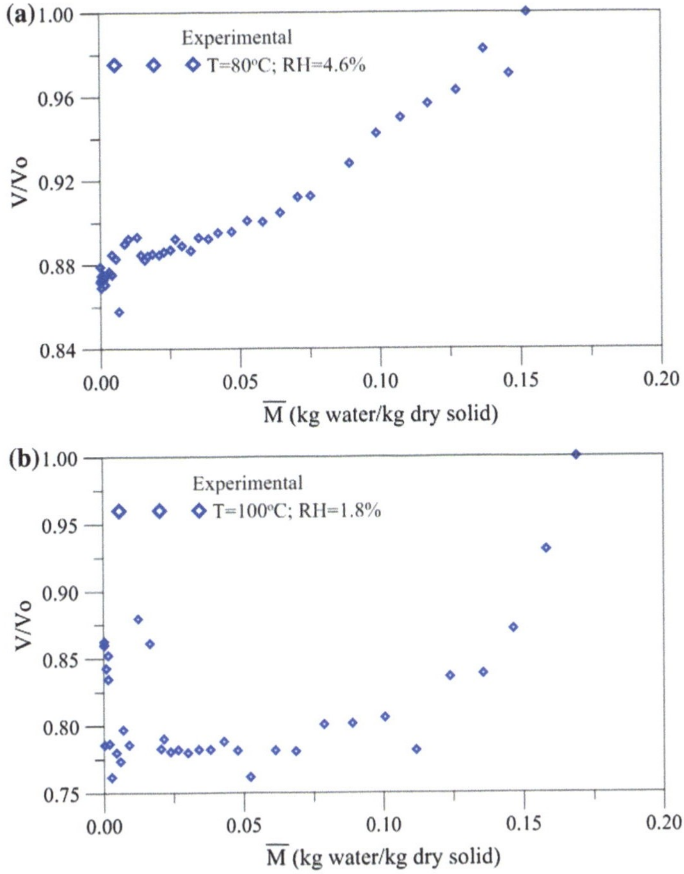

Fig. 9 Dimensionless volume of the clay brick as a function of the average moisture content. **a** T = 80 °C and **c** T = 100 °C

literature reports stress value at the external corner of approximately 0.45 MPa at the ended drying process. Obviously, this value depends on the brick material, process conformation, air and material conditions, etc. For comparison, at the equilibrium condition the maximum stresses values at the surface obtained here were σz = σx = σy = 1.105 MPa.

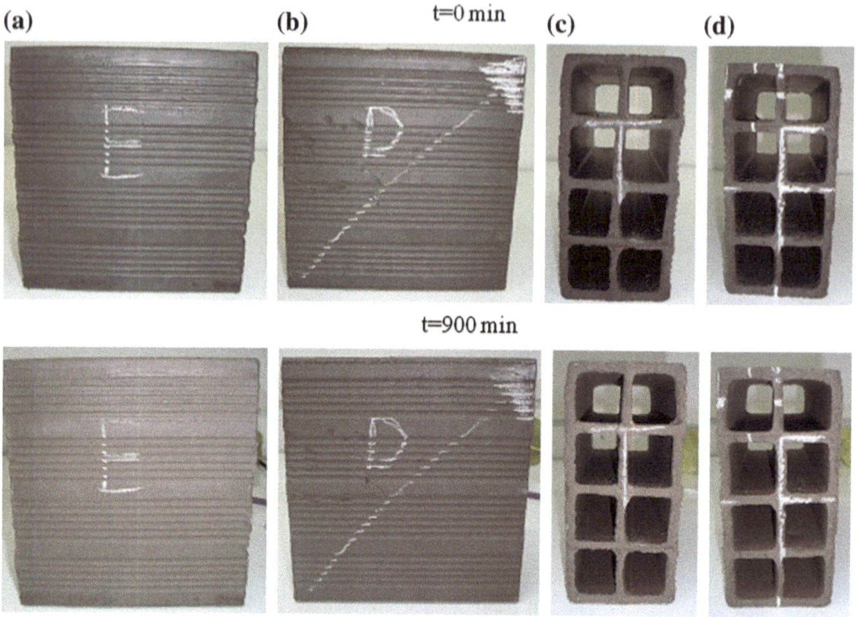

Fig. 10 View of the clay brick dried to the T = 80 °C. **a** *Left*, **b** *Right*, **c** *Back* and **d** *Front*

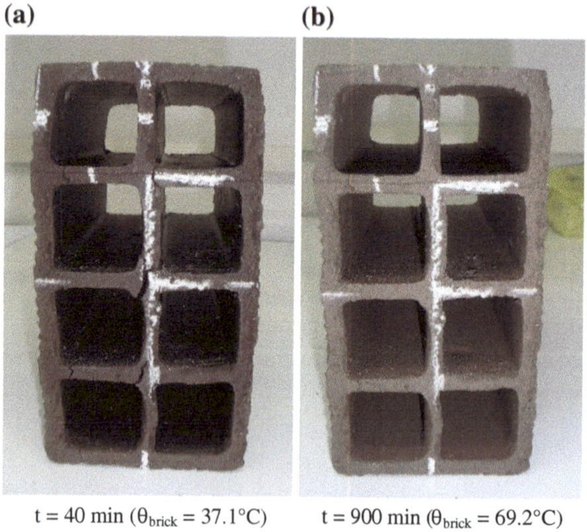

Fig. 11 View of the clay brick dried to the T = 80 °C

Fig. 12 View of the clay brick dried to the T = 90 °C

(a) (b)

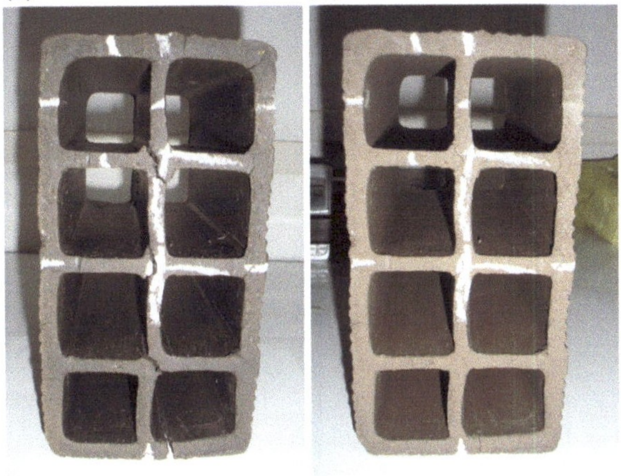

t = 20 min (θ_{brick} = 36.0°C) t = 690 min (θ_{brick} = 77.5°C)

Fig. 13 View of the clay brick dried to the T = 100 °C in the elapsed time t = 40 min. **a** *Back* and **b** *Front*

(a) (b)

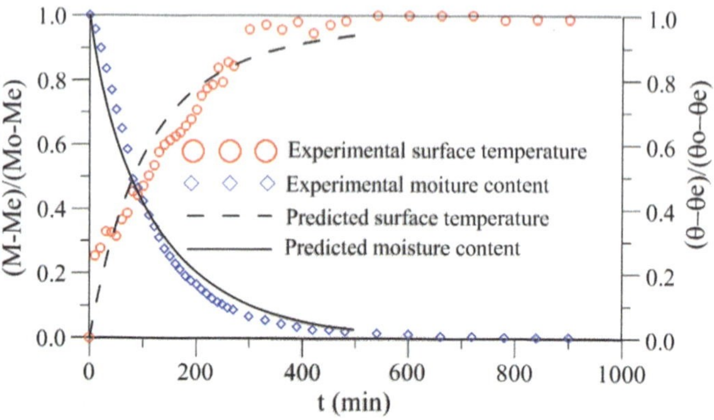

Fig. 14 Drying and heating kinetics of the industrial brick dried to T = 80 °C

Fig. 15 Dimensionless moisture content [(M − Me)/(M_o − M_e)] at the xy plane (z = R_3/2)

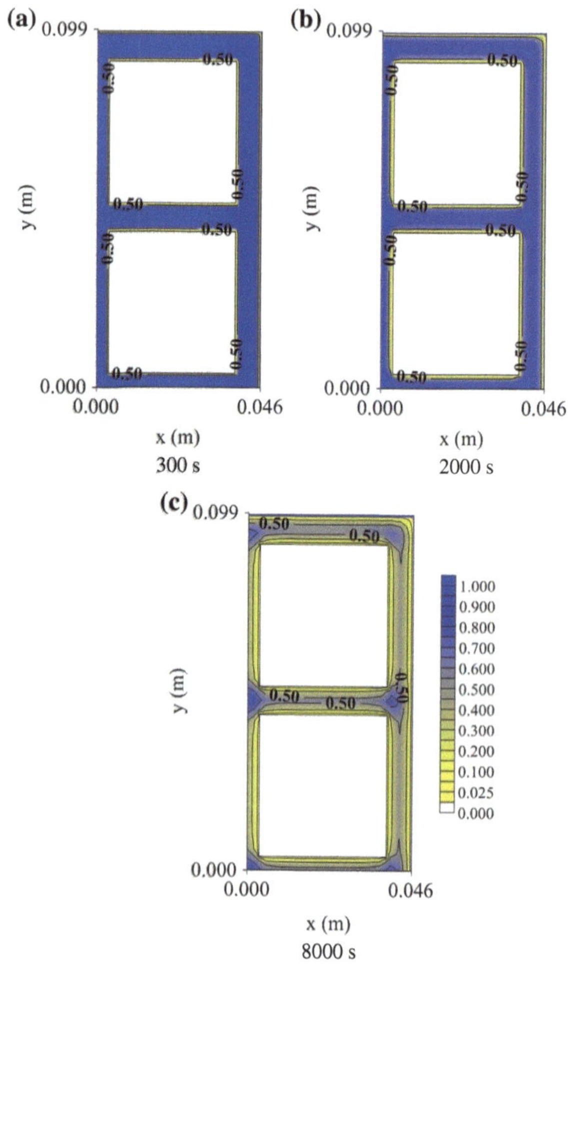

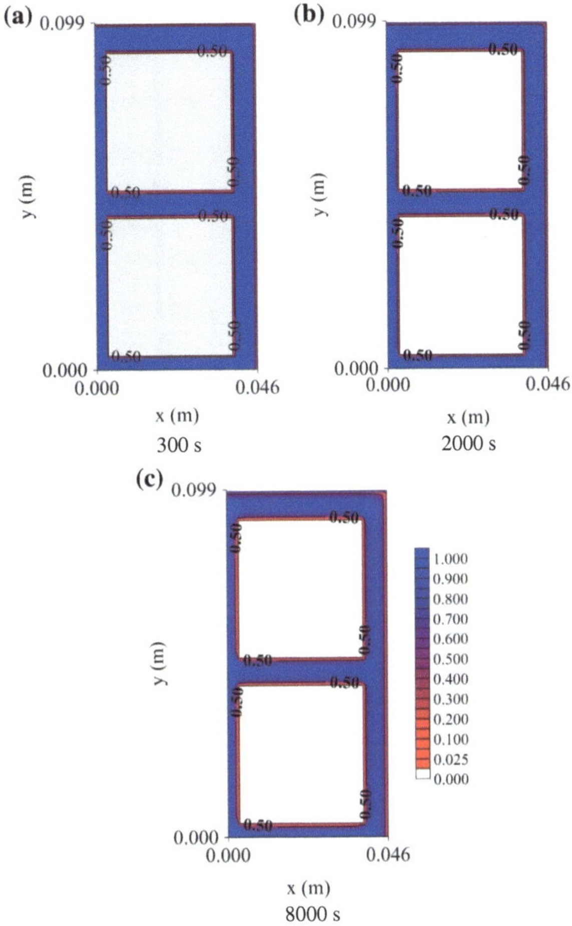

Fig. 16 Dimensionless temperature $[(\theta - \theta_e)/(\theta_o - \theta_e)]$ at the xy plane ($z = R_3/2$)

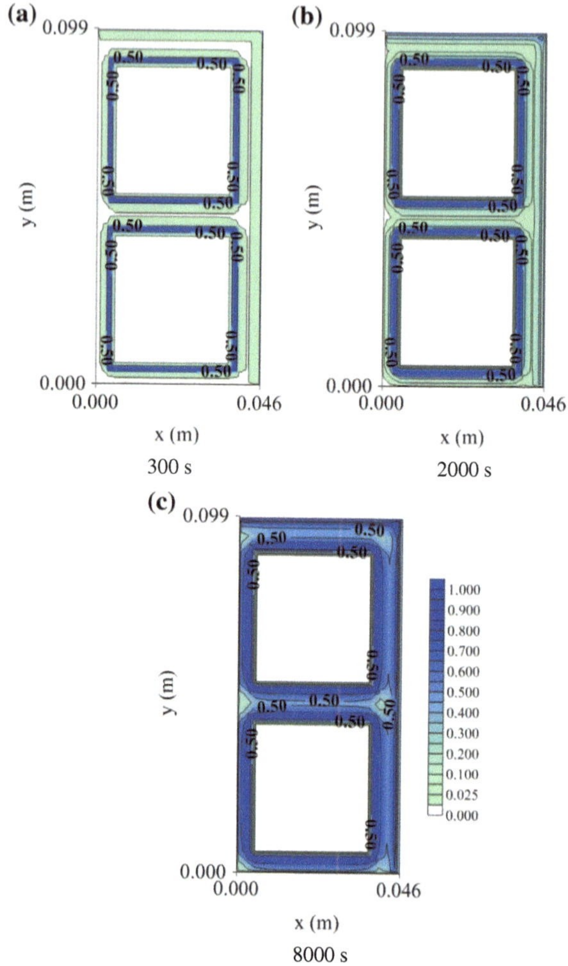

Fig. 17 Dimensionless principal stress $[(\sigma_z - \sigma_{min})/(\sigma_{max} - \sigma_{min})]$ at the xy plane ($z = R_3/2$). **a** $\sigma_{min} = 0$ and $\sigma_{max} = 1{,}103{,}471$ Pa, **b** $\sigma_{min} = 0$ and $\sigma_{max} = 1{,}103{,}473$ Pa and **c** $\sigma_{min} = 0$ and $\sigma_{max} = 1{,}103{,}473$ Pa

3 Concluding Remarks

In this chapter, a theoretical and experimental study has been presented. A three-dimensional mathematical modeling to predict heat transfer and moisture (liquid) migration in clay products with parallelepiped shape was developed, and the finite-volume method was employed to solve the governing equations. Formulation has been applied to predicted hollow brick drying process, and the stress and dimensions variations both induced by drying due to temperature and moisture changes has been investigated.

Based on the experimental results presented, the following conclusions can be summarized:

(a) The influence of temperature and relative humidity of the air as main factors in the drying rate is confirmed. As highest temperature and lowest relative humidity of the air are used in drying, heating and shrinkage rates will be higher, thus the total drying time becomes shorter.
(b) The areas near the internal and external surface (corner) of the brick present the largest rates of heat and mass transfer and largest volume variation and stresses, and so they are more susceptible areas to the cracks, fissures and deformations and even contribute to total fracture of the material;
(c) Stress due to moisture migration and high temperature can have a significant effect on the quality brick, mainly as high temperature and low relative humidity of the drying air are used.

The study indicates that the proposed modeling can be useful in:

(a) Good estimation of process time to dry the product;
(b) Reduction in energy consumption by allowing the detection of the correct processing time;
(c) Increased productivity of dried bricks, for allowing the withdrawal of the bricks from the dryer at the right time;
(d) Verification and understanding of the effect of process variables on product quality during drying and;
(e) Checking the appropriate process and dryer conditions, to improve product quality.

Acknowledgments The authors thank to FINEP, CAPES, CNPq and FACEPE (Brazilian Research Agencies) for financial support to this research, and also to the researchers for their referenced studies which helped in improving the quality of this work.

References

1. Callister Jr, W.D., Rethwisch, D.G.: Fundamentals of materials science and engineering: an integrated approach, 3rd edn. Wiley, New York (2008)
2. Callister Jr, W.D.: Materials science and engineering an introduction, 7th edn. Wiley, New York (2007)
3. Brosnan, D.A., Robinson, G.C.: Introduction to drying of ceramics. The American Ceramic Society, Westerville (2003)
4. Hasatani, M., Itaya, Y.: Modeling of strain-stress behavior for advanced drying. In: International Drying Symposium (Drying'96), vol. A, pp. 27–39. Krakow, Poland (1996)
5. Itaya, Y., Hasatani, M.: R & D needs-drying ceramics. Drying Technol. **14**(6), 1301–1313 (1996)
6. Reed, J.S.: Principles of ceramics processing. Wiley, New York (1995)
7. Sander, A., Skanki, D., Nenad, B.: Heat and mass transfer models in convective drying of clay slabs. Ceramics Inter. **29**, 641–643 (2003)
8. Ketelaars, A.A.J., Jomaa, W., Puiggali, J.R., Coumans, W.J.: Drying shrinkage and stress. In: International Drying Symposium (Drying'92), vol A, pp. 293–303 (1992)

9. Nascimento, J.J.S., Mederos, B.J.T., Belo, F.A., Lima, A.G.B.: Mass transport with shrinkage inside parallelepiped solids. Información Tecnológica **16**(1), 35–41 (2005). (In Spanish)
10. Nascimento, J.J.S., Lima, A.G.B., Teruel, B.J., Belo, F.A.: Heat and mass transfer with shrinkage during the ceramic bricks drying. Información Tecnológica **17**(6), 125–132 (2006). (In Spanish)
11. Silva, J.B., Almeida, G.S., Lima, W.C.P.B., Neves, G.A., Lima, A.G.B.: Heat and mass diffusion including shrinkage and hygrothermal stress during drying of holed ceramics bricks. Def. Diff. Forum **312–315**, 971–976 (2011)
12. Silva, W.P., Farias, V.S.O., Neves, G.A., Lima, A.G.B.: Modeling of water transport in roof tiles by removal of moisture at isothermal conditions. Heat Mass Transf. **48**(5), 809–821 (2012)
13. van der Zanden, A.J.J., Schoenmakers, A.M.E., Kerkof, P.J.A.M.: Isothermal vapour and liquid transport inside clay during drying. Drying Technol. **14**(3–4), 647–676 (1996)
14. Su, S.: Modeling of multi-phase moisture transfer and induced stress in drying clay brick. Appl. Clay Sci. **12**, 189–207 (1997)
15. van der Zanden, A.J.J.: Modelling and simulating simultaneous liquid and vapour transport in partially saturated porous materials. In: Mathematical Modeling and Numerical Techniques in Drying Technology. Marcel Dekker, Inc., New York (1997)
16. Itaya, Y., Taniguchi, S., Hasatani, M.: A Numerical study of transient deformation and stress behavior of a clay slab during drying. Drying Technol. **15**(1), 1–21 (1997)
17. Itaya, Y., Mori, S., Hasatani, M.: Effect of intermittent heating on drying-induced strain-stress of molded clay. In: International Drying Symposium (Drying'98), vol. A, pp. 240–247 (1998)
18. Augier, F., Coumans, W.J., A. Hugget, A., Kaasschieter, E.F.: On the risk of cracking in clay drying. In: International Drying Symposium (Drying'2000), CD ROM (2000)
19. Musielak, G.: Clay fracturing during drying. In: International Drying Symposium (Drying'2000), 1, CD ROM (2000)
20. Augier, F., Coumans, W.J., Hugget, A., Kaasschieter, E.F.: On the risk of cracking in clay drying. Chem. Eng. J. **86**, 133–138 (2002)
21. Banaszak, J., Kowalski, S.: Shrinkage and stress in viscoelastic cylinder in drying. In: International Drying Symposium (Drying'2000), 1, CD ROM (2000)
22. Kroes, B., Coumans, W.J., Kerkhof, P.J.A.M.: Drying behavior of deformable porous media: a mechanistic approach to clay drying. In: International Drying Symposium (Drying'96), vol. A, pp. 159–174. Krakow, Poland (1996)
23. Silva, J.B.: Simulation and Experimentation of the Drying of Drained Ceramic Bricks. Doctorate Thesis, Process Engineering, Federal University of Campina Grande, Paraíba, Brazil 174 (2009). (In Portuguese)
24. Almeida, G. S., Barbosa Fernandes, M.A.F., Ferreira Fernandes, J.N., G. Araújo Neves, G., Barbosa de Lima, W.M.P., Barbosa de Lima, A.G.B.: Drying of industrial ceramic bricks: An experimental investigation in oven. Def. Diff. Forum **353**, 116–120 (2014)
25. Silva, J.B., Almeida, G.S., Neves, G.A., Lima, W.C.P.B., Farias Neto, S.R., Lima, A.G.B.: Heat and mass transfer and volume variations during drying of industrial ceramic bricks: an experimental investigation. Def. Diff. Forum **326–328**, 267–272 (2012)
26. Almeida,G.S.: Simulation and experimentation of red ceramics drying in industrial thermal systems. Doctorate Thesis, Process Engineering, Federal University of Campina Grande, Paraíba, Brazil (2008). (In Portuguese)
27. Avelino, D.O., Nascimento, J.J.S., Lima, A.G.B., Simulation of the heat and mass transport during drying of holed ceramic bricks. In: 8o Congresso Iberoamericano de Engenharia Mecânica (CIBIM 2007), Cusco, Peru CD-ROM (2007). (In Portuguese)
28. Batista, V.R.J., Nascimento, J.S., Lima, A.G.B.: Drying and firing of common and holed ceramic bricks including dimensions variations and structural damages. Revista Eletrônica de Materiais e Processos **3**(1), 46–61 (2008). (In Portuguese)
29. Batista, V.R., Nascimento, J.J.S., Lima, A.G.B.: Drying and volumetric retraction of solid and hollow ceramic bricks: a theoretical and experimental investigation. Revista Matéria **14**(4), 1088–1100 (2009). (In Portuguese)

30. Nicolau, V.P., Lehmkuhl, W.A., Kawaguti, W.M., Dadam, A.P., Hartke, R.F., Jahn, T.G.: Numerical analysis of a continuous dryer used in red ceramic industry. In: III National Congress of Mechanical Engineering (CONEM 2004), pp. 1–10. Belém, Brazil (2004). (In Portuguese)
31. Santana, E.W.F.: Evaluation of drying and firing of ceramic plates. Master Dissertation in Science and Materials Engineering, Federal University of Campina Grande, Campina Grande, Brazil (2006). (In Portuguese)
32. Santos, G.M.: Study of thermal behavior of a tunnel kiln applied to red ceramic industry. Master Dissertation in Mechanical Engineering, Federal University of Santa Catarina, Florianópolis, Brazil (2001). (In Portuguese)
33. Nishikawa, T., Gao, T., Hibi, M., Takatsu, M., Ogawa, M.: Heat transmission during thermal shock testing of ceramics. J. Mater. Sci. **29**, 213–219 (1994)
34. Cadé, M.A., Nascimento, J.J.S., Lima, A.G.B.: Drying of holed ceramic bricks: an approach by finite-volumes. Revista Matéria **10**(3), 443–453 (2005). (In Portuguese)
35. Farias, V.S.O., Silva, W.P., Silva, C.M.D.P., Lima, A.G.B.: Three-dimensional diffusion in arbitrary domain using generalized coordinates for the boundary condition of the first kind: application in drying. Def. Diff. Forum **326–328**, 120–125 (2012)
36. Velthuis, J.F.M.: Simulation model for industrial dryers: reduction of drying times of ceramics & saving energy, In: International Drying Symposium (Drying'96), vol. B, pp. 1323–1328. Krakow, Poland (1996)
37. Recco, G.: Study to utilization of thermal energy from ceramic kiln to ceramic drying. Cerâmica Industrial **13**(3), 23–27 (2008). (In Portuguese)
38. Lehmkuhl, W.A.: Numerical and experimental analyses of a continuous dryer used in red ceramic industry. Master Dissertation in Mechanical Engineering, Federal University of de Santa Catarina, Florianópolis, Brazil (2004). (In Portuguese)
39. Hartke, R. F.: Numerical analysis of a continuous dryer used in red ceramic industry. In: 10th Brazilian Congress of Thermal Sciences and Engineering (ENCIT 2004), Rio de Janeiro, Brazil (2004). (In Portuguese)
40. Nicolau, V.P., Hartke, R.F., Jahn, T.G., Lehmkuhl, W.A.: Numerical and experimental analyses of a intermittent kiln to firing of ceramic products. In: National Congress of Mechanical Engineering (CONEM 2002), João Pessoa, pp. 1–10. Brazil (2002). (In Portuguese)
41. Almeida, G.S., Silva, J.B., Silva, C.J., Swarnakar, R., Neves, G.A., Lima, A.G.B.: Heat and mass transport in an industrial tunnel dryer: modeling and simulation applied to hollow bricks. Appl. Thermal Eng. **55**(1–2), 78–86 (2013)
42. Strumillo, C., Kudra, T.: Drying: principles, science and design. Gordon and Breach Science Publishers, New York (1986)
43. Brooker, D.B., Bakker-Arkema, F.W., Hall, C.W.: Drying and storage of grains and oilseeds. AVI Book, New York (1992)
44. Fortes, M., Okos, R.: Drying theories: their bases and limitations as applied to foods and grains . In: Mujumdar, A. (eds.) Advances in Drying. Hemisphere Publishing Corporation, Washington, vol. 1, pp. 119–154 (1980)
45. Keum, Y.T., Jeong, J.H., Auh, K.H.: Finite-element simulation of ceramic drying processes. Model. Simul. Mater. Sci. Eng. **8**, 541–556 (2000)
46. Keum, Y.T., Kim, J.H., Ghoo, B.Y.: Computer aided design of electric insulator. J. Ceramic Process. Res. **1**(1), 74–79 (2000)
47. Keum, Y.T., Oh, W.J.: Finite element simulation of a ceramic drying process considering pore shape and porosity. Model. Simul. Mater. Sci. Eng. **13**, 225–237 (2005)
48. Timoshenko, S.P., Goodier, J.N.: Theory of Elasticity. Guanabara Dois S.A., Rio de Janeiro (1980). (In portuguese)
49. Itaya, Y., Kobayashi, T., Hayakawa, K.: Three-dimensional heat and moisture transfer with viscoelastic strai-stress formation in composite food during drying. Int. J. Heat Mass Transf. **38**(7), 1173–1185 (1995)

50. Versteeg, H.K., Malalasekera, W.: An introduction to computational fluid dynamics: the finite method. Prentice Hall, London (1995)
51. Maliska, C.R.: Computational heat transfer and fluid mechanics. LTC-Livros Técnicos e Científicos Editora S.A, Rio de Janeiro (2004). (In Portuguese)
52. Patankar, S.V.: Numerical heat transfer and fluid flow. Hemisphere Publishing Corporation, New York (1980)
53. Hasatani, M., Itaya, Y.: Deformation characteristic of ceramics during drying. In: International Drying Symposium (Drying'92), vol. A, pp. 190–199. Montreal (1992a)
54. Hasatani, M., Itaya, Y.: Effect of drying process on quality control in ceramic production. In: International Drying Symposium (Drying'92), vol. B, pp. 1181–1198. Montreal (1992b)

Use of Ultrasound in the Distilled Water Pretreament and Convective Drying of Pineapple

Jefferson Luiz Gomes Corrêa, Mercedes Carolina Rasia, Jose Vicente Garcia-Perez, Antonio Mulet, João Renato de Jesus Junqueira and Juan Andres Cárcel

Abstract Convective drying is a time-consuming process. Pretreatments in distilled water assisted with ultrasound, or the application of ultrasound during convective drying itself, could improve the mass transfer. In this work, the influence of pretreatment using distilled water (20 or 40 min at 25 °C) assisted with ultrasound (55.5 W L^{-1}, 40 kHz) on convective drying kinetics (40 and 70 °C, 1 m s^{-1}) of pineapple slices (2.0 cm diameter, 0.5 cm thickness) was studied. The drying was carried out with (31 kWm^{-3}; 21.8 kHz) and without ultrasound application being the kinetics simulated using a diffusive model. The results showed that pretreatment, the use of high drying temperatures, and the application of ultrasound during drying accelerated the drying process. The effective diffusivity identified by the model allowed the quantification of the effects of each of the studied variables on the drying rate of pineapple slices.

J.L.G. Corrêa (✉) · J.R. de Jesus Junqueira
Departamento de Ciência dos Alimentos, Universidade Federal de Lavras, Lavras, Brazil
e-mail: jefferson@dca.ufla.br

J.R. de Jesus Junqueira
e-mail: jrenatojesus@hotmail.com

M.C. Rasia
Universidad Nacional de Entre Ríos, Facultad de Ciencias de la Alimentación, Concórdia, Argentina
e-mail: rasiam@fcal.uner.edu.ar

J.V. Garcia-Perez · A. Mulet · J.A. Cárcel
Depto. de Tecnología de Alimentos, Grupo de Análisis y Simulación de Procesos Agroalimentarios (ASPA), Universitat Politècnica de València, Valencia, Spain
e-mail: jogarpe4@tal.upv.es

A. Mulet
e-mail: amulet@tal.upv.es

J.A. Cárcel
e-mail: jcarcel@tal.upv.es

J.M.P.Q. Delgado and A.G. Barbosa de Lima (eds.),
Drying and Energy Technologies, Advanced Structured Materials 63,
DOI 10.1007/978-3-319-19767-8_4

List of Symbols

D_w	Effective diffusivity ($m^2\ s^{-1}$)
M_r	No dimensional moisture content
L	Half thickness of slabs (m)
S_{xy}	Standard deviation of estimation
S_y	Standard deviation of sample
T	Time (s)
W	Moisture content (kg water kg^{-1} dry solid)
% var	Explained variance

Subscripts

0	Initial condition
Eq	Equilibrium condition

1 Introduction

Drying is one of the oldest technologies used for food preservation. It is based on the reduction of water content, which decreases the availability of moisture for microbiological and enzymatic degradation reactions [35]. Convective drying is one of the most commonly used techniques for drying because of its easy employment [2, 6, 29, 46]. However, it involves the exposure of the food to high temperatures for an extended period of time, resulting in undesirable energy consumption and degradation of quality characteristics such as nutrient content, color, and mechanical properties [13, 39]. For that reason, researchers are looking for alternative methods with shorter drying times, fewer energy requirements and the potential to improve the quality of the final product [20, 33].

In this regard, high intensity ultrasound could be used to minimize both the damage incurred by drying and the energy required for this process. Ultrasound has been used during the convective drying of several food products [8, 9, 23, 25, 42, 52] or in solid-liquid pretreatments including osmotic dehydration and product immersion into distilled water [3, 21, 22]. In these cases, the effects produced by ultrasound on the internal structure of product helped with moisture elimination in a subsequent convective drying step.

The ultrasonic waves provide mechanical energy to the gas-solid or liquid-solid systems. In both cases, the use of ultrasound could reduce internal and external mass transfer resistance by its action in the solid and in the surround medium. It is important to note that the influence of the ultrasound waves on mass transfer is function of some variables as: ultrasound waves properties (mainly power and frequency); temperature, level of turbulence in the medium, medium properties (viscosity, density, surface tension and acoustic impedance) and solid properties (porosity, hardness and acoustic impedance) [11]. The energy that is actually received by the samples has to be quantified because could be the attenuated by the medium. In this sense, the mass load density has to be taken into account because more energy is received by samples when lower mass load is considered [9]. The application of ultrasound results in an additional energy available in the system. This amount of energy is more relevant in situations where the overall energy is low. On the contrary, in convective drying at high temperatures for example, the energy supplied from the air to the food is much higher than the energy supplied from the ultrasound waves. Consequently, the influence of ultrasound is reduced with temperature and could disappear at the higher ones [11, 26, 37, 43].

The agitation of the medium could disrupt the ultrasonic field, avoiding the influence the ultrasonic waves on the samples [9, 27]. This fact makes difficult the presence of ultrasonic phenomena like the microstreaming at gas-solid interfaces or the cavitation, in liquid-solid systems. Moreover, the transmission of the acoustic waves through media that present high viscosity and density is difficult. In these cases, the medium could absorb the acoustic energy, preventing the action of the ultrasonic waves [11].

Finally, the internal resistance to mass transfer in a dehydration process is related to the food structure and with the effects produced by ultrasound. Porosity can be one of the key parameters to the influence of ultrasound application in food processing [43]. In general, the higher the porosity, the softer the material and the easier the mechanical compression and expansion ("spounge effect") produced by ultrasound. Additionally, porous material presents low acoustic impedance, improving the coupling with gas system and then the transmission of ultrasonic waves. In the internal pores of solids, the diminished boundary layer could be intensified (in a gas-solid system) and cavitation could occur (in a liquid-solid system) [11, 43]. However, it has been also observed that the ultrasound application can modify the solid structure [44] of treated solids. In sum, the ultrasound application to assist dehydration processes must be assessed for each product, medium and process variables.

Fruits and vegetables dehydration supposes the reduction of moisture content [14, 16, 17, 51, 53] increasing shelf-life of products and reducing storage and transport needs. The dehydrated product could be used as a ready-to-eat product or as an ingredient in other foods such as snacks, breakfast cereals, pastries and desserts [5, 7, 20]. Besides, the dried fruits is associated with improvement in the diet and reduction in obesity [31]. They could be seen as promise food in cancer prevention due to the presence of numerous antioxidant and anti-inflammatory phytochemicals, like the phenolic compounds [30, 32]. The production of dried fruits is crescent and promissory [41]. The pineapple is a great source of vitamin C and fibers [45, 49]. According to Novagrim [40], 20 % of the world production of tropical fruits is represented by pineapple, being the greatest production from Brazil (2.12×10^6 t in 2010) and the greatest exportation is from Costa Rica (1.51×10^6 t in 2009). The pineapple drying has been studied using several techniques [15, 45, 49, 50], both with or without pretreatment [15, 21]. Although the fresh pineapple presents high content of vitamin C, the presence of this nutrient was reported to be very low in the commercial dried pineapple [36], what suggests studies for improving the drying processes.

The goal of this work was to study both the influence of ultrasonically assisted pretreatment with distilled water and the ultrasound application during convective drying, on the drying kinetics of pineapple.

2 Materials and Methods

2.1 Material

The variety of the pineapple (*Ananas comosus*) used in this work was the new hybrid 73-114, known as Golden Pineapple or MD2. It was produced in Costa Rica and purchased in a local market in Valencia, Spain. Fresh fruits with a solid soluble content ranging from 12 to 14 °Brix were chosen. The pineapples were washed and peeled. The experiments were done only with the middle part of the fruit discarding both ends parts due to the fact that the sugar content could be quite different from one extreme to another [45]. Disk shaped samples (5 mm thickness and 20 mm diameter) were obtained using a stainless steel cork borer and then blotted with tissue paper to remove surface moisture for immediate used in testing (Table 1).

The initial moisture content of the pineapple slices was determined according to the A.O.A.C. standard method 934.06 [1]. For this method, samples were placed in a vacuum oven at 70 °C and 200 mmHg until it reached a constant weight.

Table 1 Experimental conditions tested

Code	Pretreatment	Drying	
	Time (min)	Temperature (°C)	Ultrasound application
P0-D40	0	40	Non
P0-DUS40	0	40	Yes
P0-D70	0	70	Non
P0-DUS70	0	70	Yes
P20-D40	20	40	Non
P20-DUS40	20	40	Yes
P20-D70	20	70	Non
P20-DUS70	20	70	Yes
P40-D40	40	40	Non
P40-DUS40	40	40	Yes
P40-D70	40	70	Non
P40-DUS70	40	70	Yes

Each condition was tested, at least, by triplicate

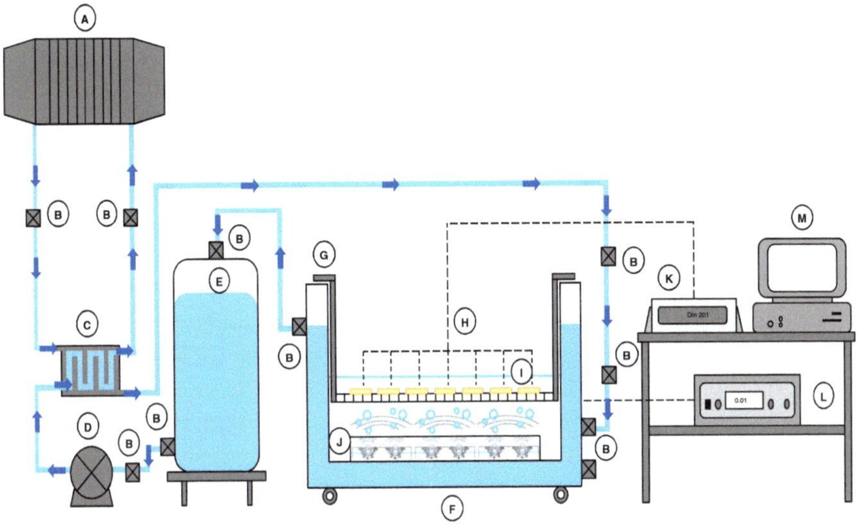

Fig. 1 Ultrasonic bath set-up. *A* chiller, *B* valve, *C* heat exchanger, *D* pump, *E* cooling reservoir, *F* ultrasonic bath, *G* sample holder, *H* thermocouples, *I* cod samples, *J* ultrasonic transducers, *K* data logger, *L* ultrasonic generator, *M* computer

2.2 Pretreatment

Pineapple samples were divided into three batches. The first batch was used directly for drying experiments (no pretreatment). The second and third batches were immersed in an ultrasonic bath (ATU S.L., Manises, Spain) with distilled water, for 20 and 40 min, respectively. The fruit to solution ratio was 1:300 (w:w). Ultrasound was applied (55.5 W L^{-1}, 40 kHz) continuously during pretreatment and the temperature was maintained at 25 ± 1 °C by recirculating a cooling fluid in the external jacket of the bath (see Fig. 1). After pretreatment, samples were blotted with tissue paper to remove surface water and, then, subjected to drying experiments.

2.3 Convective Drying

The drying experiments were carried out in the system shown in Fig. 2 [9, 25, 42]. For each run, forty samples were placed in fixed positions in the sample holder (Fig. 2) and inserted into the dryer. This holder allowed the free flow of air around each sample. The drying chamber (Fig. 2) was an aluminum cylindrical vibrating element (internal diameter 100 mm, height 310 mm and thickness 10 mm) that works as an air-borne ultrasonically activated drying chamber. The cylinder was driven by a piezoelectric composite transducer (21.8 kHz), therefore, the ultrasonic

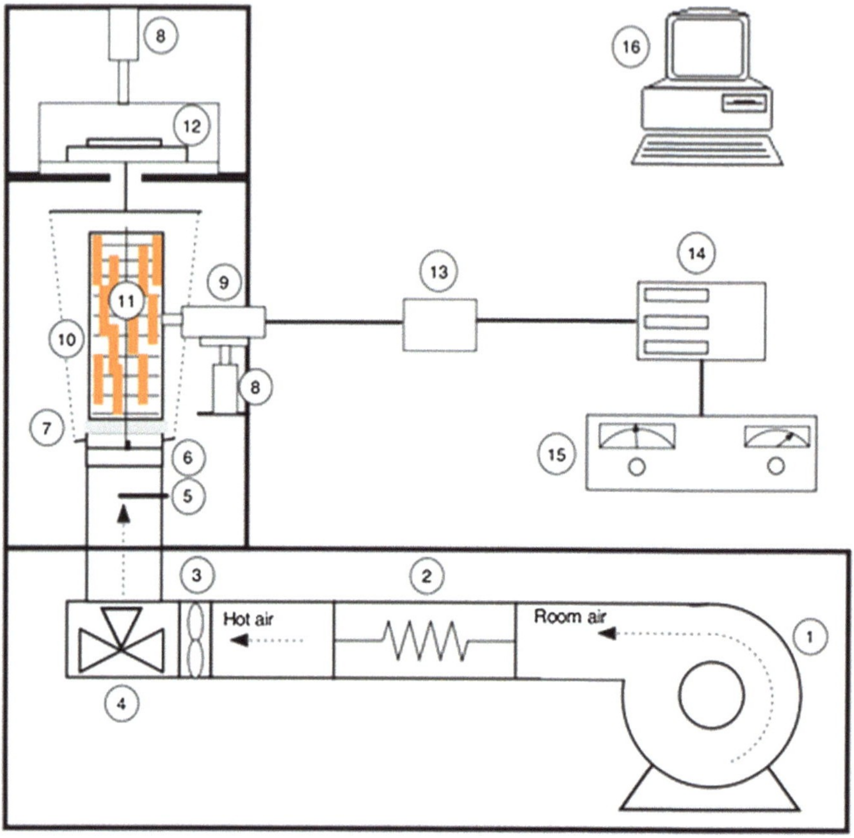

Fig. 2 Scheme of the ultrasonic assisted drier. *1* Fan, *2* heating unit, *3* anemometer, *4* 3-way valve, *5* thermocouple, *6* sample loading chamber, *7* coupling material, *8* pneumatic moving arms, *9* ultrasonic transducer, *10* vibrating cylinder, *11* sample holder, *12* balance, *13* impedance matching unit, *14* wattmeter, *15* high power ultrasonic generator, *16* PC

system was able to generate a high-intensity ultrasonic field with an average sound pressure of 154.3 dB (measured using an ultrasonic power of 31 kW m^{-3} and air stagnant conditions). The drier operated automatically and was monitored by a computer. The air velocity and temperature were controlled using a PID algorithm. The mass of the samples was measured at preset times using a scale with two pneumatic moving arms.

The drying experiments were carried out using an air velocity of 1.0 ± 0.1 ms^{-1} at 40 or 70 °C, both with and without ultrasound application (21.8 kHz, 31 kW m^{-3}) (see Table 1). The drying was terminated when samples lost 80 % of their initial weight. Each drying condition tested was performed in triplicate.

2.4 Drying Kinetics

The kinetics of convective drying was simulated using a diffusional model based on Fick's second law of diffusion [18]. Samples were assumed to behave as infinite slabs, thus mass transfer equation was reduced to a single dimension (Eq. 1).

$$\left(\frac{\partial \mathrm{W(x,t)}}{\partial \mathrm{t}}\right) = \mathrm{D_w}\left(\frac{\partial^2 \mathrm{W(x,t)}}{\partial \mathrm{x}^2}\right) \quad (1)$$

where W(x, t) is the amount of water at time t, D_w is the effective diffusivity and x is the direction of moisture flux. The initial and boundary conditions considered were: uniform initial moisture content (Eq. 2), symmetry of the solid (Eq. 3) and negligible external resistance (Eq. 4)

$$W_{(x,0)} = W_0 \quad (2)$$

$$\left.\frac{\partial W(t)}{\partial t}\right|_{z=0} = 0 \quad (3)$$

$$W_{(L,t)} = W_{eq} \quad (4)$$

Considering these assumptions, Eq. 1 can be solved analytically using separation of variables and the newly defined variable shown in Eq. 5.

$$Mr = \frac{W - W_{eq}}{W_0 - W_{eq}} \quad (5)$$

Equation 6 shows the solution of the diffusion model in terms of the average moisture content:

$$W = W_{eq} + \left(W_0 - W_{eq}\right)\left(\frac{8}{\pi^2}\sum_{i=1}^{\infty}\frac{1}{(2i+1)^2}\exp\left(-(2i+1)^2\pi^2 D_w \frac{t}{4L^2}\right)\right) \quad (6)$$

For each drying temperature, the equilibrium moisture content was determined experimentally by drying the samples for an extended period of time (>36 h). The fitting of the model was carried out to determine the effective diffusivity that minimizes the sum of the squared difference between the experimental and calculated moisture content values. For that purpose, the non-linear optimization algorithm of the Generalized Reduced Gradient (GRG), available in Microsoft Excel™ spreadsheet from MS Office 2010 (Microsoft Corporation, USA), was used. The fit of the model to the experimental data was evaluated based on the correlation coefficient (r^2) and the percentage of the explained variance (Eq. 7).

$$\%var = (1 - \frac{S_{xy}^2}{S_y^2}) \quad (7)$$

3 Results and Discussion

3.1 Experimental Drying Kinetics

The experimental drying kinetics showed the influence of each of the three variables considered in this work: pretreatment, drying temperature and the application of ultrasound during drying (Figs. 3, 4, 5, 6 and 7).

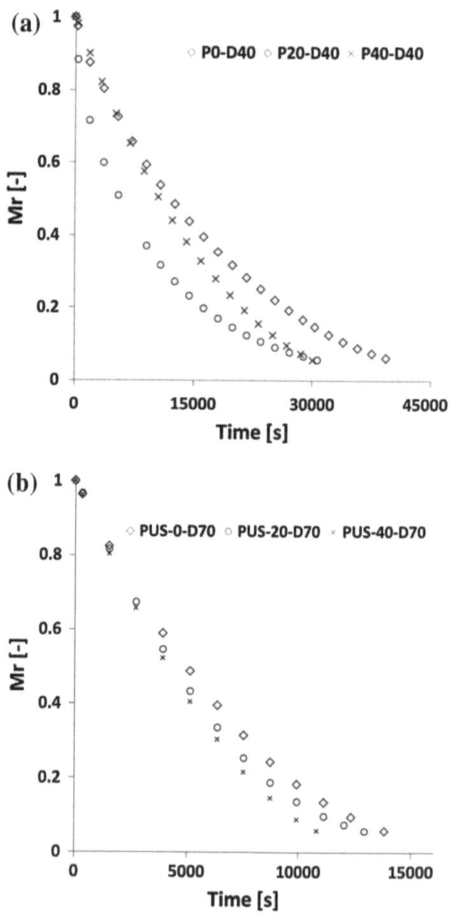

Fig. 3 Evolution of a dimensional moisture content during drying at 40 °C (**a**) and 70 °C (**b**) of pineapple samples pretreated (20 and 40 min) and non-pretreated in distilled water with ultrasound application

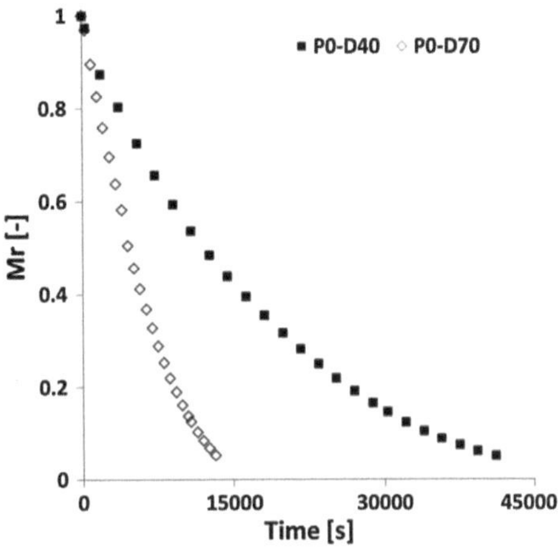

Fig. 4 Evolution of a dimensional moisture content during standard convective drying at 40 and 70 °C of pineapple samples

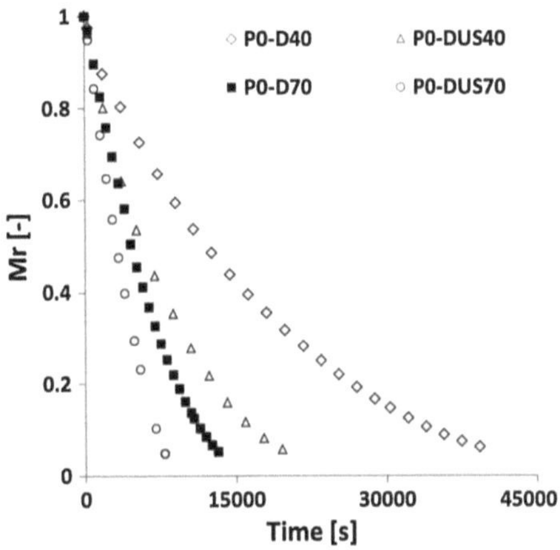

Fig. 5 Evolution of a dimensional moisture content during drying at 40 and 70 °C, with and without ultrasound application, of pineapple samples

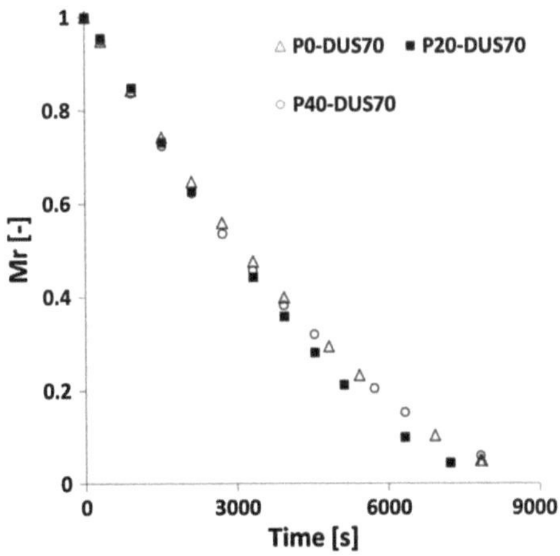

Fig. 6 Evolution of a dimensional moisture content during drying at 70 °C, with ultrasound application, of pineapple samples pretreated (20 and 40 min) and non-pretreated in distilled water with ultrasound application

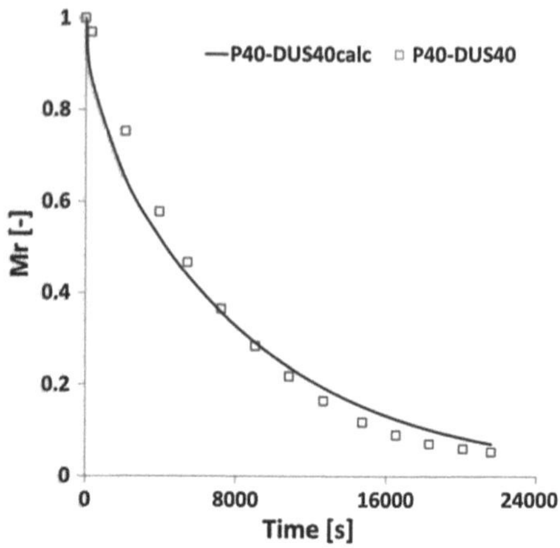

Fig. 7 Evolution of experimental and simulated a dimensional moisture content during drying at 40 °C, with ultrasound application, of pineapple samples pretreated (40 min) in distilled water with ultrasound application

3.1.1 Influence of Pretreatment

As regard to the pretreatment, the immersion of samples in distilled water with ultrasound application increased the moisture content of the samples from 5.61 ± 0.38 kg water/kg d.m. for the fresh pineapple to 7.11 ± 0.68 and 8.70 ± 0.45 kg water/kg d.m. after 20 and 40 min of pretreatment, respectively. This fact can be attributed to the higher solute content of the fruit in comparison to the distilled water, which causes the sample to uptake water by osmosis. In subsequent drying experiments, the influence of pretreatment on drying rate was observed. The drying with ultrasound application of pretreated samples was faster than in the same condition dried not pretreated samples (Fig. 3). This indicates that the pretreatment can affect the mass transfer internal resistance of samples making easier the exit of moisture from solid matrix during later drying. On one hand, the pretreatment in distilled water not only produces the gain of moisture but also the loss of soluble compounds that partially bound natural moisture content contributing to the internal resistance to mass transfer during air drying. Therefore, the pretreatment in distilled water could produce the water inside solid matrix remains more free to be eliminated. On the other hand, the application of ultrasound during pretreatment could cause structural changes in the food [34]. The mechanical stress that undergoes the samples due to the successive compression and decompression generated by the high intensity ultrasound can produce micro-channels formation [8]. Thus, Ozuna et al. [44] studied the application of ultrasound during desalting of cod. From SEM micrographs, they observed that ultrasonically desalted cod presented higher fiber swelling than conventionally desalted ones. Micrographs obtained from light microscopy also showed greater fiber and connective tissue degradation in ultrasonically treated samples. This fact can be attributed to violent microjets produced by the asymmetric implosion of bubbles near the solid surface [8]. Ozuna et al. [43] also observed structural effects produced by the applications of ultrasound during the brining of meat. All these structural effects can contribute to improve the later drying process by making easier the water movement inside the solid matrix.

It was not found a clear effect of the time of pretreatment on drying kinetics. Thus, at 40 °C, the drying of samples pretreated for 20 min was faster than the drying of those pretreated for 40 min (Fig. 3a), but, at 70 °C the opposite was observed, being faster the drying of samples pretreated for 40 min (Fig. 3b). This facts has been also observed for others authors [3, 21, 22].

3.1.2 Influence of Drying Temperature

As expected, the highest drying temperature applied, 70 °C, resulted in a faster drying of the samples (Fig. 4). Hence, the time needed to reduce the weight sample to 80 % of its initial weight at 70 °C (3.8 ± 0.2 h for P0-D70) was the 31 % of the needed at 40 °C (12.2 ± 0.9 h for P0-D40). The effect of temperature on drying rates is well reported in literature [5, 12, 28, 47]. The internal resistance to moisture transport decreased with increasing drying temperature; this was caused by an increase in the mobility of water molecules within the food product. In addition, the external resistance decreased with temperature because the water pressure in the food increases.

However, the increase of drying temperature of food products can produced the degradation of the thermo-sensitive compounds and, in general, a loss of quality. Moreover, despite of reduce the needed drying time; the increase of temperature can produce the increase of the energy consumption of the process that means higher cost and higher environmental impact.

3.1.3 Influence of Ultrasound Application During Drying

The application of ultrasound during drying clearly accelerated the drying rate of the pineapple samples as can be observed in Fig. 5. For example, the drying time decreased from 12.3 ± 0.9 h in P0-D40 experiments to 5.6 ± 0.2 h for P0-DUS40. That means a reduction of the 55 % of the needed drying time. During drying, the ultrasonic waves produce the successive compression and expansion of the samples that induces a similar effect to a sponge repeatedly squeezed and released. This phenomenon is called as "sponge effect" [19]. The forces involved in this mechanism can be higher than the surface tension which maintains the water molecules inside the capillaries of the material, creating microscopic channels [38] and facilitating the exchange of moisture. Moreover, at solid-air interfaces, ultrasound generates a micro-stirring of the gas medium and pressure variations. All these effects improve the water transfer between the solid surface and the surrounding air by the decreasing the thickness of the boundary layer and, consequently, reducing of the external resistance [9, 10, 25, 42, 48]. Therefore, the ultrasound affected both internal and external resistance that caused an increase of the drying rate.

The effect of ultrasound application on drying time was lower at 70 °C than at 40 °C. Thus, the application of ultrasound during drying at 70 °C produced a 1.6 h reduction of the drying time, from 3.8 ± 0.2 h for P0-D70 to 2.2 ± 0.2 h for P0-DUS70. Therefore, the drying assisted with ultrasound required a 42 % lower drying time than the conventional drying. This process time difference increased to 55 % when drying was carried out at 40 °C (12.3 ± 0.9 h for P0-D40 experiments and 5.6 ± 0.2 h for P0-DUS40). This observation was previously reported in literature [37]. In this sense, Gallego-Juárez [24] studying the air borne ultrasound application (155 dB) with a vibrating plate in the drying of carrots at different temperatures (60, 90 and 115 °C), observed that the effect of the ultrasonic radiation was significant at low air temperatures (60 °C) and diminished when the temperature increased. At the highest temperature tested, 115 °C, the ultrasonic effect was negligible and there were no appreciable differences in the experimental drying curves. At low temperatures, ultrasound provides mechanical energy that complements the thermal energy provided by air temperature and, jointly, they contribute to the water transport. At high air temperatures, the thermal energy is higher and masks the effect linked to the mechanical energy provided by ultrasound. Therefore, the application of ultrasound as a means of intensifying air drying processes is more efficient when a low temperature.

The drying kinetics of non-pretreated and pretreated for 20 or 40 min samples obtained at the same drying temperature and with ultrasound application were quite similar, as can be observed in Fig. 6. In these cases, the effects produced by ultrasound during drying are so important to mask the possible influence of the pretreatment on drying rate.

3.2 Modeling

The diffusion model allowed quantifying the influence of the studied variables (pretreatment, drying temperature, and ultrasound application during drying) on drying rates. It can be noted that the effective diffusivity identified by fitting the proposed model (Table 2) included every effect on drying kinetics and, therefore, also the effects of the three variables considered.

In general, the model did not provide a very accurate fit of the experimental data to the simulated data as shown the low values of the percentage of explained variance (% var) obtained (Table 2). This can be attributed to the fact that some assumptions used may greatly differ from the experimental conditions [4]. In this case, one of the boundary conditions assumed the external resistance to mass transfer as negligible compared to the internal resistance (Eq. 4). All the experiments were carried out at 1 m s^{-1}, and this air velocity may be too low to provide the air turbulence needed for totally neglect the external resistance assumed in this model. This fact can be also observed in the slight different behavior of experimental and calculated data. As can be observed in Fig. 7, the shape of the calculated data correspond with a net diffusional behavior. However the shape of the

Table 2 Effective water diffusivity identified for the convective drying of pineapple slices pretreated in distilled water with ultrasound application (55.5 W/L, 40 kHz) and dried with (37.5 kW/m^3; 21.8 kHz) or without ultrasound

Code	$D_W \times 10^{10}$ (m^2s^{-1})	r^2	% var
P0-D40	1.26 ± 0.11	0.994	96.01
P0-DUS40	2.23 ± 0.40	0.989	95.40
P0-D70	3.36 ± 0.10	0.985	92.79
P0-DUS70	4.96 ± 0.21	0.968	90.25
P20-D40	1.74 ± 0.30	0.995	97.33
P20-DUS40	2.66 ± 0.42	0.995	96.79
P20-D70	3.67 ± 0.44	0.982	92.11
P20-DUS70	6.30 ± 0.77	0.962	90.30
P40-D40	1.49 ± 0.04	0.993	95.07
P40-DUS40	2.93 ± 0.11	0.993	96.65
P40-D70	3.90 ± 0.42	0.975	90.68
P40-DUS70	5.21 ± 0.36	0.973	90.92

Average data of at least 3 replicates for each drying condition

experimental data presented a straight section at the beginning of drying that indicated the external resistance was not completely negligible compared to the internal one. However, the effective diffusivity identified by the model represented a global measurement of the kinetics of the process and was considered as adequate to quantity the influence on drying kinetics of the three factors studied.

For drying experiments carried at the same drying conditions (temperature and ultrasound application), the effective diffusivity identified for the pretreated samples was higher than that identified for the non-pretreated samples (Table 2). The highest influence of the pretreatment on effective diffusivity was observed in samples dried without ultrasound application at 40 °C. Thus, the D_W was a 38 % higher in samples P20-D40 than in P0-D40. In the case of drying experiments carried out with ultrasound application, the difference between pretreated and non-pretreated samples was not significant. The ultrasonic-assisted pretreatment may produce structural changes in pineapples that can make the drying process easier. However when ultrasound were applied during drying, the influence of the pretreatment on drying kinetics was negligible.

The time of pretreatment had no clear influence on the drying rate. Thus, the effective diffusivity identified in samples pretreated for 20 min was slightly higher than in samples pretreated for 40 min in the following conditions: drying at 40 °C without ultrasound and drying at 70 °C with ultrasound. On the contrary, the effective diffusivity of samples pretreated for 20 min was lower than for samples pretreated for 40 min when samples were dried at 40 °C with ultrasound and at 70 °C without ultrasound. However, these differences were not significant. This indicated that 20 min of pretreatment was enough to produce the structural changes in the samples that increased the kinetics. Longer pretreatments can increase the processing cost because they increase the moisture content of the samples instead of increasing the drying kinetics.

The influence of drying temperature on the drying rate was shown by the identified effective diffusivity. At the same experimental conditions, the effective diffusivity of samples dried at 70 °C was significantly higher than those dried at 40 °C. For example, the identified value for P0-D70 was 167 % higher than that identified for PUS-0-D40, and 137 % higher for P20-DUS40 than for P20-DUS70.

The application of ultrasound highly affected the values of D_w in all cases. Accordingly, the highest increase in D_w, (97 %) was observed between samples that were pretreated for 40 min and dried at 40 °C (P40-D40 and P40-DUS40). In general, the magnitude of the increase in D_w upon ultrasound application was higher for samples dried at 40 versus 70 °C, revealing a decrease in the impact of ultrasound application at high air drying temperature.

Based on the above results, it can stated that the 20 min pretreatment of pineapple in distilled water was enough for increasing the drying rate. The longer pretreatments did not provide significant reductions of drying time. The application of ultrasound during drying increased the drying rates at the temperatures tested. This fact allows obtaining the same drying kinetics but using mildest drying conditions with the potential benefits on food products quality and energy saving. Moreover, the application of ultrasound during drying masked the influence of pretreatment in drying kinetics meaning the pretreatment can be not considered in this case. However, further research is required to determine how the three variables considered here influence the quality of the final product.

4 Conclusions

The pretreatment of pineapple samples in distilled water with ultrasound accelerated the subsequent convective drying of the samples. Pretreatment with ultrasound at times longer than 20 min did not significantly affect the drying diffusivity. Ultrasound application increased the drying rate in all cases. Faster drying was achieved at higher temperatures with ultrasound application.

Acknowledgments The authors would like to thank the economical support of the Ministerio de Economía y Competitividad of Spain, the ERDF (European Regional Development Fund, DPI2012-37466-C03-03), CNPq (National Council for Scientific and Technological Development), CAPES (Coordination for the Improvement of Higher Education Personnel) and FAPEMIG (State of Minas Gerais Research Foundation) in Brazil, and Proyecto EuroTANGO 2 (Erasmus Mundus Programme).

References

1. AOAC: Official Methods of Analysis AOAC International. 18th edn. (2007)
2. Aquino, L.P., Ferrua, F.Q., Borges, S.V., Antoniassi, R., Corrêa, J.L.G., Cirillo, M.A.: Influence of pequi drying (*Caryocar brasiliense* Camb.) on the quality of the oil extracted. Ciênc Tecnol Aliment **29**(2), 354–357 (2009). doi:10.1590/S0101-20612009000200018

3. Azoubel, P.M., Baima, M.A.M., Amorim, M.R., Oliveira, S.S.B.: Effect of ultrasound on banana cv Pacovan drying kinetics. J. Food Eng. **97**(2), 194–198 (2010). doi:10.1016/j.jfoodeng.2009.10.009
4. Bon, J., Rosselló, C., Femenia, A., Eim, V., Simal, S.: Mathematical modeling of drying kinetics for apricots: influence of the external resistance to mass transfer. Dry Technol. **25**(11), 1829–1835 (2007). doi:10.1080/07373930701677918
5. Borges, S.V., Mancini, M.C., Corrêa, J.L.G., Leite, J.B.: Drying kinetics of bananas by natural convection: influence of temperature, shape, blanching and cultivar. Ciênc. Agrotec **35**(2), 368–376 (2011). doi:10.1590/S1413-70542011000200019
6. Borges, S.V., Mancini, M.C., Corrêa, J.L.G., Leite, J.: Drying of banana prata and banana d'agua by forced convection. Ciênc Tecnol Aliment **30**(3), 605–612 (2010). doi:10.1590/S0101-20612010000300006
7. Borges, S.V., Mancini, M.C., Corrêa, J.L.G., Nascimento, D.A.: Drying of pumpkin (*Cucurbita moschata* L.) slices by natural and forced convection. Ciênc Tecnol Aliment **28**, 245–251 (2008). doi:10.1590/S0101-20612008000500037
8. Cárcel, J.A., García-Pérez, J.V., Benedito, J., Mulet, A.: Food process innovation through new technologies: use of ultrasound. J. Food Eng. **110**(2), 200–207 (2012). doi:10.1016/j.jfoodeng.2011.05.038
9. Cárcel, J.A., García-Pérez, J.V., Riera, E., Mulet, A.: Improvement of convective drying of carrot by applying power ultrasound: influence of mass load density. Dry Technol. **29**(2), 174–182 (2011). doi:10.1080/07373937.2010.483032
10. Cárcel, J.A., García-Pérez, J.V., Riera, E., Mulet, A.: Influence of high-intensity ultrasound on drying kinetics of persimmon. Dry Technol. **25**(1), 185–193 (2007). doi:10.1080/07373930601161070
11. Cárcel, J.A., García-Pérez, J.V., Riera, E., Rosselló, C., Mulet, A.: Drying assisted by ultrasound (Chapter 8) In: Tsotsas E, Mujumdar AS (eds.) Modern Drying Technology, vol. 5, pp. 237–278. Process intensification. Wiley New York (2014). ISBN 978-3-527-31560-4
12. Chua, K.J., Mujumdar, A.S., Hawlader, M.N.A., Chou, S.K., Ho, J.C.: Convective drying of agricultural products. Effect of continuous and stepwise change in drying air temperature. Dry Technol. **19**(8), 1949–1960 (2001). doi:10.1081/DRT-100107282
13. Corrêa, J.L.G., Braga, A.M.P., Hochheim, M., Silva, M.A.: The Influence of ethanol on the convective drying of unripe, ripe, and overripe bananas. Dry Technol. **30**(8), 817–826 (2012). doi:10.1080/07373937.2012.667469
14. Corrêa, J.L.G., Cacciatore, F.A., Da Silva, Z.E., Arakaki, T.: Osmotic dehydration of West Indian cherry (*Malpighia emarginata* D.C.)—mass transfer kinetics. Rev. Ciênc Agron **39**(3), 403–409 (2008)
15. Corrêa, J.L.G., Dev, S.R.S., Gariepy, Y., Raghavan, G.S.V.: Drying of pineapple by microwave-vacuum with osmotic pretreatment. Dry Technol. **29**(13), 1556–1561 (2011). doi:10.1080/07373937.2011.582558
16. Corrêa, J.L.G., Ernesto, D.B., Alves, J.G.L.F., Andrade, R.S.: Optimisation of vacuum pulse osmotic dehydration of blanched pumpkin. Int. J. Food Sci. Technol. **49**(9), 2008–2014 (2014). doi:10.1111/ijfs.12502
17. Corrêa, J.L.G., Pereira, L.M., Vieira, G.S., Hubinger, M.D.: Mass transfer kinetics of pulsed vacuum osmotic dehydration of guavas. J. Food Eng. **96**(4), 498–504 (2010). doi:10.1016/j.jfoodeng.2009.08.032
18. Crank, J.: The mathematics of diffusion, 2nd edn. Claredon Press, Oxford (1975)
19. De la Fuente, S., Riera, E., Acosta, V.M., Blanco, A., Gallego-Juárez, J.A.: Food drying process by power ultrasound. Ultrasonics **44**, e523–e527 (2006). doi:10.1016/j.ultras.2006.05.181
20. Fante, C., Corrêa, J., Natividade, M., Lima, J., Lima, L.: Drying of plums (Prunus sp, c.v Gulfblaze) treated with KCl in the field and subjected to pulsed vacuum osmotic dehydration. Int. J. Food Sci. Technol. **46**(5), 1080–1085 (2011). doi:10.1111/j.1365-2621.2011.02619.x

21. Fernandes, F.A.N., Linhares, F.E., Rodrigues, S.: Ultrasound as pre-treatment for drying of pineapple. Ultrason. Sonochem. **15**(6), 1049–1054 (2008). doi:10.1016/j.ultsonch.2008.03.009
22. Fernandes, F.A.N., Rodrigues, S.: Ultrasound as pre-treatment for drying of fruits: dehydration of banana. J. Food Eng. **82**(2), 261–267 (2007). doi:10.1016/j.jfoodeng.2007.02.032
23. Frias, J., Peñas, E., Ullate, M., Vidal-Valverde, C.: Influence of drying by convective air dryer or power ultrasound on the vitamin C and β-carotene content of carrots. J. Agric. Food Chem. **58**(19), 10539–10544 (2010). doi:10.1021/jf102797y
24. Gallego-Juárez, J.A.: Some applications of air-borne power ultrasound to food processing. In: Povey, M.J.W., Mason, T.J. (eds.) Ultrasound in Food Processing, pp. 127–143. Chapman & Hall, London (1998). ISBN 978-0-7514-0429-6
25. García-Pérez, J.V., Ozuna, C., Ortuño, C., Cárcel, J.A., Mulet, A.: Modeling ultrasonically assisted convective drying of eggplant. Dry Technol. **29**(13), 1499–1509 (2011). doi:10.1080/07373937.2011.576321
26. García-Pérez, J.V., Rosselló, C., Cárcel, J.A., De la Fuente, S., Mulet, A.: Effect of air temperature on convective drying assisted by high power ultrasound. Defect Diffus. Forum 258–260 (2006). doi:10.4028/www.scientific.net/DDF.258-260.563
27. García-Pérez, J.V., Cárcel, J.A., Benedicto, J., Mulet, A.: Power ultrasound mass transfer enhancement in food drying. Food Bioprod. Process **85**, 247–254 (2007). doi:10.1205/fbp07010
28. Isquierdo, E.P., Borém, F.M., Andrade, E.T., Corrêa, J.L.G., Oliveira, P.D., Alves, G.E.: Drying kinetics and quality of natural coffee. Trans. ASABE **56**(3), 1003–1010 (2013)
29. Jangam, S.V.: An overview of recent developments and some R&D challenges related to drying of foods. Dry Technol. **29**(12), 1343–1357 (2011). doi:10.1080/07373937.2011.594378
30. Kamiloglu, S., Pasli, A.A., Ozcelik, B., Capanoglu, E.: Evaluating the in vitro bioaccessibility of phenolics and antioxidant activity during consumption of dried fruits with nuts. LWT Food Sci. Technol. **56**(2), 284–289 (2014). doi:10.1016/j.lwt.2013.11.040
31. Keast, D.R., O'Neil, C.E., Jones, J.M.: Dried fruit consumption is associated with improved diet quality and reduced obesity in US adults: National Health and Nutrition Examination Survey, 1999–2004. Nutr. Res. **31**(6), 460–467 (2011). doi:10.1016/j.nutres.2011.05.009
32. Kundu, J.K., Chun, K.S.: The promise of dried fruits in cancer chemoprevention. Asian Pac. J. Cancer Prev. **15**, 3343–3352 (2014). doi:10.7314/APJCP.2014.15.8.3343
33. Kumar, C., Karim, M.A., Joardder, M.U.H.: Intermittent drying of food products: a critical review. J. Food Eng. **121**, 48–57 (2014). doi:10.1016/j.jfoodeng.2013.08.014
34. Lee, H., Feng, H.: Effect of power ultrasound on food quality. In: Feng, H., Barbosa-Canovas, G., Weiss, J. (eds.) Ultrasound Technologies for Food and Bioprocessing. Springer, Berlin (2011). ISBN 978-1-4419-7471-6
35. Marques, G.R., Borges, S.V., Botrel, D.A., Costa, J.M.G., Silva, E.K., Corrêa, J.L.G.: Spray drying of green corn pulp. Dry Technol. **32**(7), 861–868 (2014). doi:10.1080/07373937.2013.873452
36. Megías-Pérez, R., Santos, J.G., Soria, A.C., Villamiel, M., Montilla, A.: Survey of quality indicators in commercial dehydrated fruits. Food Chem. **150**, 41–48 (2014). doi:10.1016/j.foodchem.2013.10.141
37. Mulet, A., Cárcel, J.A., Sanjuán, N., Bon, J.: New food drying technologies: use of ultrasound. Food Sci. Technol. Int. **9**, 215–221 (2003)
38. Muralidhara, H.S., Ensminger, D., Putnam, A.: Acoustic dewatering and drying (low and high frequency): state of the art review. Dry Technol. **3**(4), 529–566 (1985). doi:10.1080/07373938508916296
39. Niamnuy, C., Devahastin, S., Soponronnarit, S.: Some recent advances in microstructural modification and monitoring of foods during drying: a review. J. Food Eng. **123**, 148–156 (2014). doi:10.1016/j.jfoodeng.2013.08.026
40. Novagrim: Fresh Fruit Importer and Vegetable Supplier since 1999 (2014). http://www.novagrim.com/Pages/2000_2011_pineapple_statistics_EN.aspx. Accessed 08 Dec 2014

41. Nutfruit: International Nut and dried fruit "Global Statistical Review 2007–2012" (2012). http://www.nutfruit.org/glob-stat-review-2011-2012_70816.pdf. Accessed 08 Dec 2014
42. Ozuna, C., Cárcel, J.A., García-Pérez, J.V., Mulet, A.: Improvement of water transport mechanisms during potato drying by applying ultrasound. J. Sci. Food Agric. **91**(14), 2511–2517 (2011). doi:10.1002/jsfa.4344
43. Ozuna, C., Álvarez-Arenas, T.G., Riega, E., Cárcel, J.A., García-Pérez, J.V.: Influence of material structure on air-borne ultrasonic application in drying. Ultrason. Sonochem. **21**(3), 1235–1243 (2014). doi:10.1016/j.ultsonch.2013.12.015
44. Ozuna, C., Puig, A., Garcia-Perez, J.V., Cárcel, J.A.: Ultrasonically enhanced desalting of cod (Gadus morhua). Mass transport kinetics and structural changes. LWT Food Sci. Technol. **59** (1), 130–137 (2014). doi:10.1016/j.lwt.2014.05.062
45. Ramallo, L.A., Mascheroni, R.H.: Quality evaluation of pineapple fruit during drying process. Food Bioprod. Proc. **90**(2), 275–283 (2012). doi:10.1016/j.fbp.2011.06.001
46. Rodríguez, J., Mulet, A., Bon, J.: Influence of high-intensity ultrasound on drying kinetics in fixed beds of high porosity. J. Food Eng. **127**, 93–102 (2014). doi:10.1016/j.jfoodeng.2013.12.002
47. Sadeghi, M., Mirzabeigi Kesbi, O., Mireei, S.A.: Mass transfer characteristics during convective, microwave and combined microwave-convective drying of lemon slices. J. Sci. Food Agric. **93**(3), 471–478 (2013). doi:10.1002/jsfa.5786
48. Santacatalina, J.V., Rodríguez, O., Simal, S., Cárcel, J.A., Mulet, A., García-Pérez, J.V.: Ultrasonically enhanced low-temperature drying of apple: influence on drying kinetics and antioxidant potential. J. Food Eng. **138**, 35–44 (2014). doi:10.1016/j.jfoodeng.2014.04.003
49. Santos, P.H.S., Silva, M.A.: Kinetics of L-ascorbic acid degradation in pineapple drying under ethanolic atmosphere. Dry Technol. **27**(9), 947–954 (2009). doi:10.1080/07373930902901950
50. Silva, M.A., Corrêa, J.L.G.: Academic research on drying in Brazil 1970–2003. Dry Technol. **23**(7), 1345–1359 (2005). doi:10.1081/DRT-200063473
51. Silva, M.A., Corrêa, J.L.G., Da Silva, Z.E.: Application of inverse methods in the osmotic dehydration of acerola. Int. J. Food Sci. Technol. **45**(12), 2477–2484 (2010). doi:10.1111/j.1365-2621.2010.02378.x
52. Soria, A.C., Villamiel, M.: Effect of ultrasound on the technological properties and bioactivity of food: a review. Trends Food Sci. Technol. **21**(7), 323–331 (2010). doi:10.1016/j.tifs.2010.04.003
53. Viana, A.D., Corrêa, J.L.G., Justus, A.: Optimisation of the pulsed vacuum osmotic dehydration of cladodes of fodder palm. Int. J. Food Sci. Technol. **49**(3), 726–732 (2014). doi:10.1111/ijfs.12357

Drying Process in Electromagnetic Fields

A.G. Barbosa de Lima, J.M.P.Q. Delgado, E.G. Silva, S.R. de Farias Neto, J.P.S. Santos and W.M.P. Barbosa de Lima

Abstract In this chapter a comprehensive study of the phenomena involved in the dehydration process of wet porous solids through the use of electromagnetic waves with the aim to improve process strategies has been presented. Electromagnetic waves penetrate into the material and heat it volumetrically, due to the interaction of the electric field with water molecules. Fundamental aspect of electromagnetic fields, and microwave, sun and infrared drying is reviewed and the effect of drying method on the drying rate, heating rate and product quality have been discussed along with different applications. A general mathematical modeling based on the diffusion theory (mass and heat conservation equations), Maxwell's equations, Lambert's law and Poynting vector theorem that predicts the volumetric heating and drying behavior of bioproduct was derived and the importance of this procedure of analysis on the drying process optimization were also discussed.

A.G. Barbosa de Lima (✉) · J.P.S. Santos · W.M.P.B. de Lima
Department of Mechanical Engineering, Federal University of Campina Grande, Av. Aprígio Veloso, 882, Bodocongó, 58429-900 Campina Grande, PB, Brazil
e-mail: antonio.gilson@ufcg.edu.br

J.P.S. Santos
e-mail: jplitys@bol.com.br

W.M.P.B. de Lima
e-mail: wan_magno@hotmail.com

J.M.P.Q. Delgado
CONSTRUCT-LFC, Civil Engineering Department, Faculty of Engineering (FEUP), University of Porto, Porto, Portugal
e-mail: jdelgado@fe.up.pt

E.G. Silva
Department of Physics, CCT, State University of Paraíba (UEPB), 58429-500 Campina Grande, PB, Brazil
e-mail: dinasansil@yahoo.com.br

S.R. de Farias Neto
Department of Chemical Engineering, Federal University of Campina Grande (UFCG), 58429-900 Campina Grande, PB, Brazil
e-mail: fariasn@deq.ufcg.edu.br

J.M.P.Q. Delgado and A.G. Barbosa de Lima (eds.),
Drying and Energy Technologies, Advanced Structured Materials 63,
DOI 10.1007/978-3-319-19767-8_5

Keywords Drying · Heating · Microwave · Infrared · Sun · Modeling · Product quality

1 Electromagnetic Fields: Fundamental Concepts

What are electromagnetic waves? Light, microwaves, X-rays, and TV and radio transmissions are all kinds of electromagnetic waves. They are all the same kind of wavy disturbance that repeats itself over a distance called the wavelength. Electromagnetic waves carry energy as they travel through empty space. The rate of energy transport is associated with both the spatial variation of the electric field E and the magnetic field B amplitudes, being perpendicular to both the electric and magnetic fields and in the direction of propagation of the wave. An electromagnetic wave can be created by accelerating charges; moving charges back and forth will produce oscillating electric and magnetic fields, and these travels at the speed of light. So, electromagnetic wave is a transverse wave and its energy is stored in the electric and magnetic fields (Fig. 1).

Something interesting about electromagnetic waves is that can travel through a vacuum or a material medium. Other waves, such as sound waves, require a medium to travel.

The different kinds of electromagnetic waves, such as light and microwaves, form the electromagnetic spectrum. Here is a diagram of the electromagnetic spectrum that has appeared in the published literature (Fig. 2).

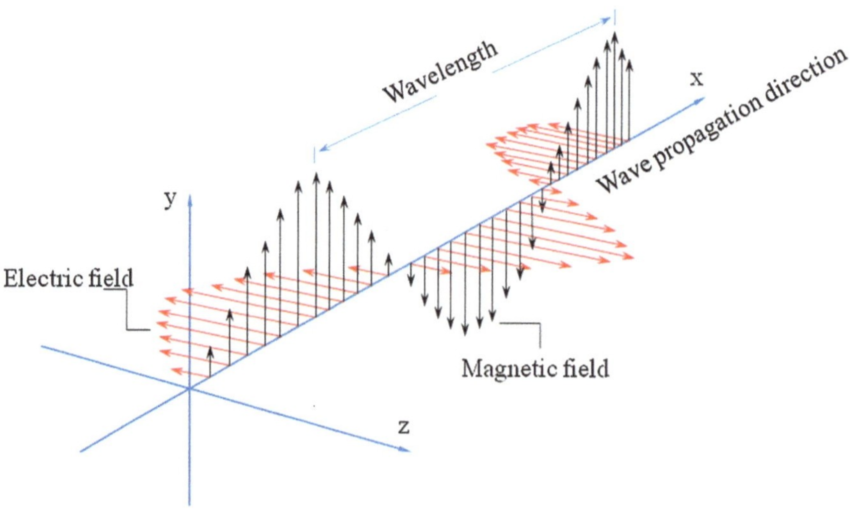

Fig. 1 An electromagnetic wave and the oscillating electric and magnetic fields

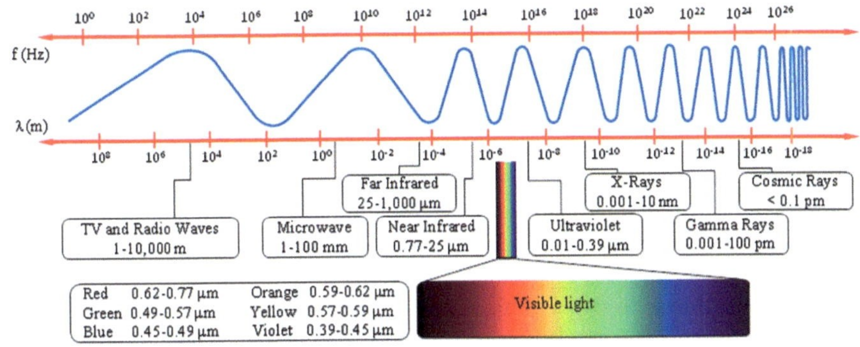

Fig. 2 The electromagnetic radiation spectrum

There are a number of basic properties of electromagnetic waves; among them are speed, wavelength and frequency. Wavelength is the distance between a given point on one cycle and the same point on the next cycle as shown. Frequency is the number of times that a particular point on the wave moves up and down in a given time (normally a second). All electromagnetic waves, regardless of their frequency, travel through a vacuum at the same speed, the speed of light.

The wave speed of an electromagnetic wave in a dielectric medium is given by $v = 1/(\varepsilon_0\mu_0)^{1/2}$, where $\varepsilon_0 = 8.854 \times 10^{-12}\,\mathrm{F/m}$ and $\mu_0 = 4\pi \times 10^{-7}\,\mathrm{H/m}$ are the permittivity and permeability of the dielectric of free space, respectively. One other mathematical expression that can be used to calculate wave velocity in a dielectric medium is

$$v = \frac{c}{\sqrt{\varepsilon'}} \tag{1}$$

where c is the speed of light in air and ε' is the dielectric constant of the material through which the wave is propagated.

Relationship between frequency and wavelength is given as follows:

$$f = \frac{v}{\lambda} \tag{2}$$

When an electromagnetic wave propagates through a material its frequency remain the same, but the wavelength varies.

2 Drying in Electromagnetic Fields

2.1 Basic Concepts in Drying

The presence of water into the materials significantly increases the bulk volume as well as mass of it leading to difficulties in handling and transport. Further, in biological materials such as fruits, vegetables and grains, the moisture increases the water activity, enabling microbiological spoilage and degradation of the material. Thus, removal of the moisture in wet materials is crucial. Depending of the material nature moisture removal can be made by osmotic dehydration, pressing and drying.

Drying of wet porous solids is very complex transient process involving simultaneous heat and mass transport and often both phase change into the solid and dimension variations due to moisture removal (termed as shrinkage) and heating (termed as volumetric expansion). The goal is to remove moisture from the materials with minimum effect over their mechanical, physical and chemical characteristics and low energy cost.

There are different drying techniques being used today. They includes convective, freeze, vacuum, fixed bed, spray, rotary, pneumatic, flash, fluidized bed, spouted bed, ohmic, heat pump, superheated steam, ultrasound, solar, infrared and microwave drying and combinations thereof. Energy consumption and other processing cost, and quality and shelf life of dried products are main parameters for correct selection of a drying technique and dryer type.

2.2 *Dielectric Drying*

2.2.1 Microwave Drying

Microwaves are non-ionizing electromagnetic waves whose frequency range from 300 MHz to 300 GHz. This makes microwave wavelengths varies from 1 mm to 1 m. In general, domestic microwave appliances operate at a frequency of 2.45 GHz, while industrial microwave systems operate at frequencies of 915 MHz and 2.45 GHz [1, 2].

Materials interact with electromagnetic fields in different form. Thus, they can be classified into four classes: Conductors (for example, metals), Insulators (For example, ceramics and air), Dielectrics (for example, fruits, vegetables, grains, wood, and many others wet porous materials) and Magnetic compounds (for example, ferrites) [3]. Further, based on the microwave absorption characteristics, materials can be classified into three categories [2]: (i) absorbers or high dielectric loss materials which are strong absorbers of microwave (ii) transparent or low dielectric loss materials where microwave energy passes through the material with little attenuation and (iii) opaque or conductors which reflect the microwaves. Hence, knowledge of dielectric properties becomes necessary to differentiate the materials behavior under microwave action. Dielectric material, for example, it has both free and bound charges which are accelerated by electric fields. Free charges may move translationally and bound charges can move rotationally and vibrationally [4].

Microwaves are not forms of heat but rather forms of energy that are converted in heat through their interaction with materials. They are not absorbed by the material due to its electronic or atomic polarization. There are many mechanisms for this energy conversion. The main are: ionic conduction and dipolar rotation. The latter one is regarding with the dipolar molecules. Many molecules, such as water, are dipolar in nature, i.e., they possess an asymmetric charge center. Other molecules may become "induced dipoles" because of the stresses caused by the electric field.

Dipolar rotation: When an oscillating electric field is incident on the water molecules, the permanently polarized dipolar molecules try to realign in the direction of the electric field. Due to the high frequency the electric field, this realignment occurs at a million times per second and causes internal friction of molecules resulting in the volumetric heating of the material [2, 5]. Thus, the microwave energy is converted into kinetic energy, which makes water molecules vibrate strongly causing friction and internal heating [6].

Ionic conduction: Microwave heating might also occur due to the oscillatory migration of ions in the materials which generates heat in the presence of a high frequency oscillating electric field [7].

Dipole polarization is significant at frequencies above 1 GHz while ionic losses are predominant at frequencies below 1 GHz [2, 8]. Further, water into the materials can be found in free and bound state. Under the action of an electric field, the polar molecules of water in a free state orient more freely than those of bound water.

Heating of materials by using microwave and dielectric is volumetric (the energy is transmitted directly to the bulk of the wet porous material), thus, being more uniform compared to the many traditional heating methods, like conduction and convection, for example (Fig. 3). However, they produce non-uniform temperature distribution inside the material, hence; some regions of the material are heated quickly, while others regions get heated with less intensity. This is one of the major problems associated with the microwave heating [2]. The heating occurs nearly instantaneously, generally some seconds or minutes.

The part of the electromagnetic wave energy incident on the surface of the material which was absorbed falls in intensity as the wave passes through it. In the absence of magnetic loss (dielectric material) literature has reported that microwave power converted into heat per volume unit of drying material due to electric loss (dipolar rotation) is equal to the product of electric conductivity and the square of the electric field intensity as [2, 9–12]:

$$P_{\text{microwave}} = \frac{1}{2}\sigma\left|\vec{E}\right|^2 \tag{3}$$

Since that $\sigma = \omega\varepsilon''$ and the electric field intensity drops with the penetration deep it can be written as follows:

$$P_{\text{microwave}} = \frac{1}{2}\omega\varepsilon'' E_{\text{o}}^2 \exp(-2\psi\vec{r}\cdot\vec{n}) \tag{4}$$

Further, the angular frequency $\omega = 2\pi f$, so, in term of the frequency Eq. (3) can be written as follows:

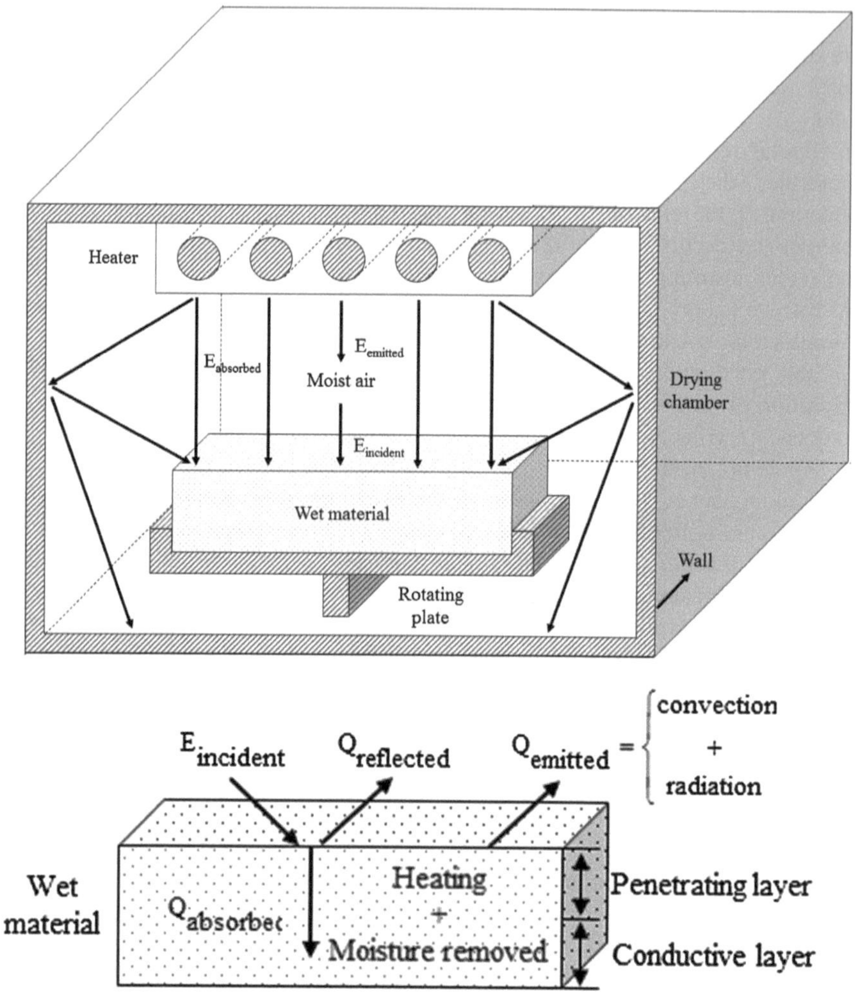

Fig. 3 Scheme of the action of electromagnetic waves in a moist media

$$P_{\text{microwave}} = \pi f \varepsilon'' \left|\vec{E}\right|^2 \tag{5}$$

Or yet

$$P_{\text{microwave}} = \pi f \varepsilon' \tan(\delta) \left|\vec{E}\right|^2 \tag{6}$$

where E is the electric field strength within the material, in volts per unit distance, f is the frequency, ε' is the dielectric constant, or permittivity, tan δ is the loss tangent or dissipation factor, and ε'' is the loss factor. In the case of uniform field within the wet porous material, during drying, for example, in small samples, $|\vec{E}| = E_0 = \sqrt{2}E_{rms}$, where E_0 corresponds to the peak magnitude of the electric field and E_{rms} represent the root mean square value (rms value) of the electric field. Putting these results into Eq. 5 is obtained [13–22]:

$$P_{microwave} = 2\pi f \varepsilon'' E_{rms}^2 \tag{7}$$

Since that water is the main absorber of the microwave energy, the permittivity depends on the temperature and moisture content. Based on this characteristic, the following equation has been reported in the literature [11, 23]:

$$P_{microwave} = P_{surface,microwave} \frac{M(\vec{r},t)}{M_o} \varepsilon''(M,T) \exp(-2\psi \vec{r} \cdot \vec{n}) \tag{8}$$

where $P_{surface,\ microwave}$ is a parameter to be determined experimentally through the measurement of energy used for evaporation of some amount of free water in a specific microwave dryer; ε'' represent the dielectric dissipation factor; ψ is the microwave attenuation coefficient; $\vec{r}$ and $\vec{n}$ represent the position vector and outward unit normal vector, respectively.

By analyzing Eq. (8) it can be seen that the microwave heat source is null for the dry solid, i.e., as the moisture content is equal to zero.

The relative dielectric constant and relative dielectric loss respectively, are given as follows:

$$\varepsilon_r' = \frac{\varepsilon'}{\varepsilon_0} \tag{9}$$

and

$$\varepsilon_r'' = \frac{\varepsilon''}{\varepsilon_0} \tag{10}$$

where ε' is the dielectric constant, which signifies the ability to store electric energy and ε'' is the dielectric loss, which signifies the ability to convert electric energy into heat, and ε_0 represents the permittivity of free space.

The relative dielectric constant expresses the degree to which an electric field may build up within a material when a dielectric field is applied to the material.

The loss tangent is a measure of how much of that electric field will be converted into heat. A further examination of Eq. (7) reveals that E and f are functions of the equipment, whereas ε', ε'', and $\tan\delta$ are factors related to the material that is heated. Another important point is that as frequency f is changed is necessary to increase the electric field strength E, in order to maintain a particular power level P.

There are many factors which affect microwave heating and the most important of them are the dielectric properties and penetration depth.

Dielectric properties are affected for different parameters such as: moisture content, density, temperature, frequency, electric conductivity, specific heat, and penetration depth.

The amount of free or bound moisture in a substance greatly affects its dielectric constant since water has a high dielectric constant. The higher the moisture content, usually the higher is the dielectric constant. Since drying is concerned with the removal of water or a solvent, it is interesting to note that as these liquids are removed the dielectric loss decreases and hence, the material heats less well. In many cases, this leads to self-limitation of the heating as the material becomes relatively transparent at low moisture content.

Table 1 reports different equations for the parameters ε' and ε'' as a function of the moisture content and/or temperature, which has been used in the literature.

The relative value of the dielectric constant as applied to composites which incorporates different materials can be given as follows

$$\varepsilon' = v_1\varepsilon_1' + v_2\varepsilon_2' + \cdots + v_n\varepsilon_n' \tag{11}$$

where v_n represents the volume fraction of the material [32].

The parameters affecting the depth of the field into the material are the wavelength, the dielectric constant, and the loss factor, as expressed in Eq. (12).

$$\hat{D} = \frac{\lambda_o\sqrt{2}}{2\pi}\left[\varepsilon_r'\left(\sqrt{1+\left(\frac{\varepsilon_r''}{\varepsilon_r'}\right)^2} - 1\right)\right]^{-1/2} \tag{12}$$

where $\hat{D}$ is the penetration depth at which the available power in the material drops to a value of 37 % ($1/e$) from its value at the surface and λ_o is the free space wavelength. Relationship between penetration deep and microwave attenuation coefficient is

$$\hat{D} = \frac{1}{\psi} \tag{13}$$

From these equations, it is obvious that materials with high dielectric constants and loss factors will have smaller depths of penetration than those with lower values. It is also apparent that the depth of penetration is greatly affected by the wavelength (and hence the frequency) of the applied field.

This demonstrates the very clear superiority of dielectric heating of very large materials with substantial dielectric properties.

A peculiarity of this type of heating is the unusual temperature gradients that may be generated. This is due to a number of factors. First, unless an auxiliary heat form is applied, the air in the system remains cold. Hence, the surface will be cooler than a zone somewhat below the surface. This is especially true in a drying system in which evaporative cooling of the surface will occur.

Table 1 Predictive equations for dielectric properties of different materials relative to the permittivity at the vacuum

Equations	Material	Range	Source
$\varepsilon_r' = 2.14 - 0.104T + 0.808M'$ $\varepsilon_r'' = 3.09 - 0.0638T + 0.213M'$	Fruits and vegetables	25–60 °C	Bingol et al. [24]
$\varepsilon_r' = 81.79 - 0.299T$	Potatoes (*Solanum tuberosum*)	32.61–94.03 °C and 70.78–201.43 °C	Pandit and Prasad [25]
$\varepsilon_r'' = 22.6 - 0.378T + 0.00293T^2$	Sodium alginate gel–Standard gel	23–77 °C	Lin et al. [26]
$\varepsilon_r' = 117.6 - 0.026T + 1.8M - 0.073\rho$	Potato and carrot	10–80 °C	Jia et al. [27]
$\varepsilon_r'' = 12.2 + 0.0014T + 0.096M + 0.0039\rho$	Carrot, carrot embedded with teflon	25–90 °C $0.5 < M < 6.0$ kg/kg	Sanga et al. [28]
$\varepsilon_r' = -0.0351M^4 + 1.7369M^3 - 20.293M^2 + 81.206M + 2.5$ $\varepsilon_r'' = 0.0529M^6 - 1.1602M^5 + 9.4946M^4 - 34.61M^3 + 47.75M^2 + 3.8164M$	Potato	$0.5 < M < 5.5$ kg/kg	Malafronte et al. [12]
$\varepsilon_r' = 71.06 - 0.052T - 0.00083T^2$ $\varepsilon_r'' = 20.95 - 0.25T + 0.0014T^2$	Fresh pear	20–100 °C	Arballo et al. [29]
$\varepsilon_r' = 0.158M' + 1.74$ $\varepsilon_r'' = 0.0332M' - 0.0154$	Fruits: apple, banana, kiwi and strawberry	$M' < 21$ %	Kristiawan et al. [30]
$\varepsilon_r' = 1.14M' - 18$ $\varepsilon_r'' = 0.608M' - 12$		21 % $< M' <$ 50 %	Kristiawan et al. [30]
$\varepsilon_r' = 0.056M' + 2.95$ $\varepsilon_r'' = 0.024M' + 0.18$	Vegetables: broccoli and onion	$M' < 39$ % $M' < 29$ %	Kristiawan et al. [30]
$\varepsilon_r' = 1.31M' - 37.2$ $\varepsilon_r'' = 0.65M' - 18$		32 % $< M' <$ 65 % 29 % $< M' <$ 65 %	Kristiawan et al. [30]
$\varepsilon_r' = 38.57 + 0.1255T + 0.4546M' - 14.54A - 0.0037M'T + 0.07327AT$ $\varepsilon_r'' = 17.72 - 0.4519T + 0.001382T^2 - 0.07448M' + 22.93A - 13.44A^2 + 0.002206M'T + 0.1505AT$	Overall		Sipahioglu and Barringer [31]
$\varepsilon_r' = 22.12 + 0.2379T + 0.5532M' - 0.0005134T^2 - 0.003866M'T$ $\varepsilon_r'' = 33.41 - 0.4415T + 0.001400T^2 - 0.1746M' + 1.438A + 0.001578M'T + 0.2289AT$	Fruits: Apple, banana, corn, cucumber and pear	T: 20–100 °C; A (Ash): 0.26–0.72 ° C; M': 76.67–95.89 %	Sipahioglu and Barringer [31]
$\varepsilon_r' = -243.6 + 1.342T + 4.593M' - 426.9A + 376.5A^2 - 0.01415M'T - 0.3151AT$ $\varepsilon_r'' = -100.02 - 0.1611T + 0.001415T^2 + 2.429M' - 378.9A + 316.2A^2$	Vegetables: carrot, broccoli, garlic, potato, radish, spinach, turnip, yam, parsley, parsnip	T: 5–130 °C; M': 57.30–95.82 %; A: 0.53–1.56 %	Sipahioglu and Barringer [31]

T Temperature (°C), M moisture content (decimal d.b.), M' moisture content (%, w.b.), A ash content (%, w.b.), ρ product density (kg/m^3)

Another circumstance concerns the depth of penetration as it relates to the size of the piece that is heated. If the piece is several times larger than the depth of penetration, then the temperature gradient will resemble conventional gradients, with a cooler interior and a warmer exterior. However, if the piece is small in comparison with the penetration depth, for example only one or two times greater, then there may be a focused accumulation of the electromagnetic field in the center of the piece due to the multiple passes of the waves and internal reflections. In this case, the center may be the hottest place, and in fact, if it is overheated, the center may burn whereas the surface remains cool.

Recently, microwaves are being widely used to heating instead of the conventional heating methods mainly for shorter and effective processing at different scales [33]. Thus, microwave drying is now a new and promising thermal processing technique developed in recent years.

Heating by using microwave present many advantages such as, process speed is increased, uniform heating may occur throughout the material, efficiency of energy conversion, better and more rapid process control, faster start-up and shut-down times, floor space requirements are usually less, selective heating may occur, product quality may be improved and desirable chemical and physical effects may result. Besides, thermal-insulating material has low thermal conductivity and consequently, heat penetration efficiency by conduction is very small. Thus, only the water that reaches the surface can acquire enough energy to be evaporated, and the drying rate is very slow. In this case microwave drying is recommended [34–40].

Microwave energy acts selectively on the water molecules which are weakly bounded to the solid substrate making it to be evaporated quickly without a temperature increase. As only the water is heated no energy is lost in warming the atmosphere surrounding the material during drying process [41].

Despite of these advantages drying with microwaves or dielectrics alone can be very expensive in terms of both equipment and operating costs. Thus, this technique traditionally has been used in combination with other conventional drying methods. Further, microwave energy has been applied in the falling rate period or at low moisture content to finish drying process [42].

Theoretical and experimental studies of microwave drying of different material has been reported in the literature, such as Torriga et al. [43] (sugar cubes); Martínez-Monzó et al. [44] (orange slices); Ahrens and Habeger [45] (paper); Martín et al. [35] (apple slices); Berteli and Marsaioli Jr. [46] (Pasta); Beke et al. [47] (beetroot); Rodriguez et al. [48] (pasta); Dong et al. [49] (brown rice); McLoughlin et al. [50] (pharmaceuticals products); Askari et al. [42] (apples and tomato slices); Eman-Djomeh et al. [39] (strawberry); Hasna [32] (composite); Raghavan et al. [36] (carrot and apple cubes); Lescano et al. [51] (okara); Nordin et al. [52] (pitaya); Shalmashi [53] (caffeine); McMinn et al. (pharmaceutical products) [54]; Wang and Chen [55] (pharmaceutical products); Itaya et al. [56] (clay slab); Andrés et al. [57] (cassava slices), and Bihercz and Kurjak [58] (potato, onion, paprika and parsley root).

Interesting applications of microwave energy have recently not only found in drying but also in other food processing such as cooking, pasteurization, extraction, chemical reaction, thawing, blanching, tempering and sterilization [44, 59–62].

Despite of the importance, it is well known that microwave radiations cause serious health problems such as loss of appetite, irritation, discomfort, fatigue and headaches (microwave syndrome) [63].

2.2.2 Infrared Drying

All forms of matter (solid, liquid and gas) emit radiation in all directions by virtue of its temperature (thermal radiation) its transport does not require the presence of any matter medium. However, what is the origin and the nature of this transport? The physical mechanism of the radiation is not yet completely understood. Two theories are proposed: (i) propagation of a collection of particles termed photons or quanta, and (ii) propagation of electromagnetic waves. Here attention is given to latter theory [64, 65].

Infrared drying is a drying technique that uses infrared electromagnetic waves as energy source. The wavelengths used are in range of 0.76–400 μm [66]. Traditionally, the wavelength of infrared has been used for classify it as near infrared (0.75–3.00 μm), intermediate infrared (3.00–25.00 μm), and far infrared (25–100 μm) [67].

Infrared radiation can be generated electrically, by passing an electric current through an electric resistance (joule effect) (for example, metal sheart radiant rods, quartz tubes, and quartz lamps), or by a burning gas that heats a plate (for example, metal or refractory ceramic), which then emits the radiation [67, 68]. Obviously electric energy is expensive if compared with energy generated directly by combustion of a gas, but drying using electric energy source is clean (no pollutant emissions). Thermal energy emitted by a surface occurs in a range of wavelength and temperature.

The emissive power varies with the temperature and wavelength. The total emissive power per unit area increases when the temperature increases and varies with the surface characteristic of the emitter material according to following equation:

$$E_{\text{emitted,infrared}} = \varepsilon_r \sigma_r T^4 \ (\text{W/m}^2) \tag{14}$$

where ε_r represent the emissivity of the material surface, $\sigma_r = 5.670 \times 10^{-8}\ \text{W/m}^2\text{K}^4$ is the Stefan–Boltzmann constant, and T is the absolute surface temperature.

The nature of the emission coefficient varies with roughness and color of the surface, type and temperature of the material, which depend of method of fabrication, thermal cycling, and chemical reaction with its environment [69].

The infrared thermal energy received by matter can be absorbed, reflected, or transmitted (Fig. 3). The absorptivity $\hat{\alpha}$ is a property that determines the fraction of the absorbed energy by a surface. The reflectivity $\hat{\rho}$ represents the fraction of the incident energy reflected by a surface. The transmissivity $\hat{\tau}$ represent the fraction of the incident energy that is transmitted through the material. The values of these parameters are dependent of the state and nature of the material and wavelength of the radiation. Under this viewpoint, material can be classified in transparent, semitransparent and opaque. For better understanding, a material such as glass or water, which is semitransparent at short wavelengths, becomes opaque at longer wavelength. Relations between these three parameters are given below:

$$\hat{\rho} + \hat{\alpha} + \hat{\tau} = 1 \tag{15}$$

Hence knowledge of one property can be used to calculate the value of other.

Most solid are considered opaque ($\hat{\tau} = 0$); the energy absorbed by a solid is converted into heat. This energy penetrates the surface of the material and causes vibrations of molecules which create the thermal effect [66]. So, the part of energy absorbed is the only effective part to be used for drying of wet porous solid.

The IR energy absorbed at the surface of the solid is expressed in terms of the incident energy as follows:

$$P_{\text{infrared}} = \hat{\alpha} E_{\text{incident,infrared}} \tag{16}$$

Because of the IR wavelength the penetration depth of infrared radiation is relatively small (≈several mm), IR is mainly used in drying process of thin material such as adhesives, paper, textiles coatings, and paints [66–68].

The radiation absorption of water in the liquid phase is considerably high in the near and intermediate infrared range.

The radiation heat flux (IR) that penetrates into the material can be determined by Bouguer's law as follows:

$$P = P_{\text{infrared}} \exp(-\int_0^z a dz) \tag{17}$$

where a refers to the absorption coefficient [70–72].

For material constituted by two or more different material, absorption coefficient is determinate by taking a weighted average of absorption coefficient of each material.

According to Noboa and Seyed-Yagoobi [71], because IR heating/drying is non-contact mode to heat transfer becomes it very attractive for drying of coated surfaces and pre-heating and drying of uncoated paper and grades [72].

IR can be used continuously or intermittently during drying process but it is preferred as long duration as possible in the first stage of drying where drying rate is higher [73]. However, IR radiation and temperature must be carefully controlled to minimize product quality degradation.

Theoretical and experimental studies on IR drying of different material can be found in the reported literature. For example, Torriga et al. [43] (Sugar Cubes), Seyed-Yagoobi et al. [70] (paper sheet); Allanic et al. [74] (water based varnishes); Seyed-Yagoobi and Noboa [72] (uncoated paper); Dontigny et al. [75] (graphite slurry).

2.3 Sun Drying

The electromagnetic waves emitted by the sun shows a wide range of wavelengths. According to the capability of ionizing atoms in radiation-absorbing matter solar radiation can be classified in ionizing radiation (X-rays and gamma-rays) and non-ionizing radiation (Ultraviolet radiation, visible light, and infrared radiation). Fortunately, the highly injurious ionizing radiation does not penetrate the earth's atmosphere, thus solar radiation is composed by only non-ionizing radiation.

The sun emits almost all of its energy in a range of wavelengths from about 0.1 and 3 μm. Most of this energy is in the visible light region. Figure 4 illustrates the solar spectrum in terms of energy intensity.

Solar radiation are partially depleted and attenuated as it passes through the atmosphere layers due to absorption, scattering, and reflection, which occurs in the

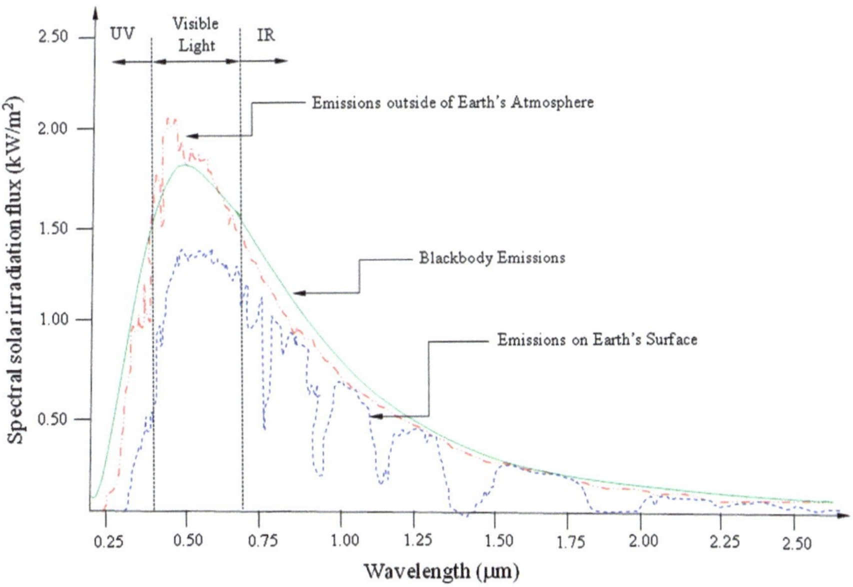

Fig. 4 The spectral solar irradiation (thermal radiation)

stratosphere in its thin layer of ozone, and in the troposphere, because the cloud formations and weather conditions [76–81].

Open-air sun drying (natural convection or conventional solar drying) has been used for centuries to dry textiles, clay, and plants, marine, agricultural and food products and other materials containing moisture, and the like. During sun drying, the energy is transferred by convection from the surrounding air and by direct and diffuse radiation from the sun to the surface of the solid. One part of this energy propagates to the interior of the product (direct absorption) causing increase in the temperature and phase change of liquid water in vapor, and the remaining is used to evaporate liquid water at the surface of the solid, which is removed by air natural convection, and lost to ambient by convection and radiation [82].

Among the advantages of solar drying one may cite a free, nonpolluting, renewable, abundant energy source that cannot be monopolized [83]. Further, the use of sun and solar drying can strongly reduce the use of fossil fuels. Despite of these advantages, solar radiation daily presents periodic character, i.e., incident radiation varies with the time and is weather dependent; a problem that can be minimized by storing or converting part of the energy gained during radiation periods using electric devices (for example, photovoltaic cells). Further, it involves placing the production on the ground, mats or rocks in the sun [84]. Thus, expensive investment and high energy cost are expected to made indirect solar dryer. Further, in conventional solar drying (natural or sun drying), the product is unprotected from rain, wind-borne dirt and dust, enzymatic reactions, infestation by insects, rodents and other animals, which may be seriously degraded by resulting in loss of dried product quality.

During sun drying, due to hygroscopic nature of the agricultural products they can either be dried or rewetted depending of the temperature and relative humidity of the environment air. Rewetting can occurs by water condensation or by vapor diffusion caused by osmotic or capillary forces [85]. Non-uniform and insufficient drying also leads to reduced shelf life of the crop. Then, knowledge of cost effective of direct solar dryers (sun drying) is very useful, and helps the socio-economic development of the developing tropical and sub-tropical countries, especially in Asia and Africa, since the initial cost of construction of indirect solar dryers is the main reason why these dryers are not used extensively in crop drying [84, 86].

Theoretical and experimental studies of sun drying has been reported in the literature. For example, Gibriel et al. [87] (apricots and grapes); Sablani et al. [88] (fish sardines); Sakin et al. [89] (biscuit baking); Sankat and Mujaffar [84] (shark fillets).

3 Drying Mathematical Modeling

An appropriate knowledge of the transport mechanisms involved on drying of wet porous materials in electromagnetic fields is the most important aspect in heat and mass transfer analysis, especially in heat sensitive materials such as agricultural products (fruits, vegetables and grains), and for equipment design, process optimization, and increasing the efficiency. Mathematical modeling of electromagnetic heating of wet porous materials is very complex. The reason for this is the number of coupled partial differential equations, describing the physical problems of electromagnetism, and heat and mass transfer, which have to be solved simultaneously. Modeling of electromagnetic heating involves the use of electromagnetic equations and energy equations to predict the temperature distribution as well as power absorption inside the products, while moisture removal is established by mass (liquid and/or vapor) diffusion equation.

Some of the recent developments in the drying mathematical modeling of drying using electromagnetic waves are based on the lumped (mass) and distributed (mass and heat) analysis. Mass transfer obeys diffusion theory or strong simplifications of it. Heating models can be classified as: (a) models that consider isolation of the heat equation with a nonlinear source term assuming the electric field amplitude as constant, (b) models that involve the coupling between the electric field amplitude and temperature including both steady-state solutions and the initial heating, (c) models that control thermal runaway and (d) numerical models that elaborate on the microwave heating process [18, 21, 90–93].

Generally, it has been verified that the published model are limited mostly to special cases or to very similar ones, where they have been applied successfully.

Anyway, the fundamental mathematical equations to describe drying process are developed on the basis of the mass and heat balances in a differential volume into the wet porous material. In general is assumed that the solid is composed by moisture and dry substrate, and that all the phases are continuously distributed in space within the material.

The following general transient diffusion equation (short form) applied for any coordinate system can be used:

$$\frac{\partial(\zeta\Phi)}{\partial t} = \nabla.(\Gamma^{\Phi}\nabla\Phi) + S^{\Phi} \tag{18}$$

Equation (18) has three terms: the term of the left side refers to accumulative term; the first term of the right side corresponds to diffusive term, and the second refers to the volumetric generation term S^{Φ}. In this equation, ζ and Γ^{Φ} represent transport properties, Φ is the potential known and t is the time.

For a well-posed formulation, the following initial and convective boundary conditions can be used:

$$\Phi = \Phi_o \text{ at } t = 0 \tag{19}$$

$$-\Gamma^{\Phi}\nabla\Phi\,|_{surface} = h(\Phi - \Phi_e)|_{surface} + S_T \quad \text{at } t > 0\,\text{s} \tag{20}$$

where h represent convective transfer coefficient, Φ_e is the potential known at the equilibrium condition ($t \to \infty$), and S_T is a term incorporating any others considered physical effects at the surface of the solid.

For mass transport $\zeta = \rho$ (solid density), $\Gamma^{\Phi} = \rho D$ (D is the diffusion coefficient), $\Phi = M$ (moisture content, dry basis), $h = h_{m}$ (convective mass transfer coefficient), and $S^{\Phi} = S_{T} = 0$. For heat transfer $\Gamma^{\Phi} = k$ (thermal conductivity), $\zeta = \rho c_{p}$ (c_{p} represent specific heat), $\Phi = \theta$ (solid temperature), $h = h_{c}$ (convective heat transfer coefficient). The term S^{Φ} represents the part of the incident energy at the surface of the solid (absorbed energy) which will be used to heating of the solid and evaporating the liquid water into the one. For example, in microwave drying $S^{\Phi} = P_{\text{microwave}}$ or $S^{\Phi} = P_{\text{infrared}}$ in infrared drying.

The term S_{T} for microwave drying and infrared drying are given as follows:

$$S_{T} = -\dot{m}_{w}h_{fg} + \varepsilon_{r}\sigma_{r}\left(T^{4}_{\text{surface}} - T^{4}_{\text{surrounding}}\right) \tag{21}$$

$$S_{T} = -\dot{m}_{w}h_{fg} + \varepsilon_{r}\sigma_{r}\left(T^{4}_{\text{surface}} - T^{4}_{\text{surrounding}}\right) + P_{\text{surface infrared}} \tag{22}$$

being ε_{r} is the emissivity, σ_{r} is the Stefan-Boltzmann constant, $\dot{m}_{w}$ and h_{fg} represent the evaporated mass flow rate and latent heat of water vaporization, respectively. So, the first term of the right side of Eqs. 21 and 22 takes into account heat either lost by evaporation or gained by vapor condensation. When the source term $S^{\Phi} = 0$, the model reduces to the conventional convective drying.

The average temperature and average moisture content can be calculated by classical mode.

$$\bar{\Phi} = \frac{1}{V}\int_{v} \Phi dV \tag{23}$$

where V is the volume of the solid.

We can notice that despite or even though mathematical formulation presented here takes account only the heat and mass transfer phenomena. One more complete formulation must consider the stress treatment, and liquid and vapor fluxes inside the material [94, 95].

4 Concluding Remarks

Electromagnetic heating has vast applications in the field of food processing such as baking, cooking, pasteurization, extraction, chemical reaction, thawing, blanching, tempering, sterilization, pasteurization, drying and preservation of food materials. This chapter focuses on the electromagnetic heating of wet porous materials, together with information about diverse drying techniques with particular reference to microwave, infrared and sun drying processes. Comparison between electromagnetic waves and conventional (hot air) drying methods has been done. The comparison highlights that the microwave drying offers several advantages over conventional heating methods and is more efficiently methods for moisture removal among them. This result is an important achievement in advancing the knowledge of how microwaves interact with complicated anisotropic materials such as food-stuffs. Thus, the potential value of electromagnetic waves assistance in drying of foods at different scale applications is explored.

A more general transient mathematical modeling to describe heat and mass transfer during drying in presence of electromagnetic fields has been present in details. Mass transfer is based on the diffusion theory while heat transfer is modeled on the basis of the Fourier's law, Maxwell's equations, Lambert's law equations and Poynting Vector Theorem. There are two important issues that have been discussed herein for coupling the electromagnetic and thermal processes: the electromagnetic wave energy required for heating and evaporating the water inside the wet materials, which depends on the physical, mechanical and electrical properties, and shape of them, and the moisture removal mechanism. The factors affecting the dielectric property of food material were also discussed.

The temperature distribution effect into the material during microwave heating is primarily discussed through the penetration depth parameter, shape, size and position of a particular material relative to electromagnetic wave direction. These mechanisms are extremely important in controlling the drying process. Nevertheless, knowledge of the interrelationship among electromagnetic theory, dielectric response, models, experiments, and applications of electromagnetic heating to materials processing, should improve our fundamental understanding of drying processes in electromagnetic fields and complement experimental work toward better optimization of the current processes.

Acknowledgments The authors would like to express their thanks to CNPq (Conselho Nacional de Desenvolvimento Científico e Tecnológico, Brazil), CAPES (Coordenação de Aperfeiçoamento de Pessoal de Nível Superior, Brazil), and FINEP (Financiadora de Estudos e Projetos, Brazil) for supporting this work; to the authors of the references in this paper that helped in our understanding of this complex subject, and to the Editors by the opportunity given to present our research in this book.

References

1. Datta, A.K., Anantheswaran, R.C.: Handbook of Microwave Technology for Food Applications. Marcel Dekker Inc., New York (2000)
2. Chandrasekaran, S., Ramanathan, S., Basak, T.: Microwave food processing—a review. Food. Res. Int. **52**, 243–261 (2013)
3. Schiffmann, R.F.: Microwave and Dielectric Drying. In: Mujumdar, A.S. (ed.) Handbook of Industrial Drying, pp. 285–305. CRC Press, Boca Raton (2007)
4. Schimidt, P.S., Bergman, T.L., Pearce, J.A.: Heat and mass transfer considerations in dielectrically—enhanced drying. In: 8th International Drying Symposium (IDS 92), Montreal, Canada, Part A, pp. 137–160 (1992)
5. Garcia, A., Iglesias, O., Roques, M., Bueno, J.L.: Microwave drying of agar gels: kinetic parameters. In: 8th International Drying Symposium (IDS'92), Montreal, Canada, Part A, pp. 595–606 (1992)

6. Sutar, P.P., Prasad, S., Sutar, N., Thorat, B.N.: Effect of microwave power density and pressure on selected quality parameters of dehydrated carrots. In: 16th International Drying Symposium (IDS'2008), Hyderabad, India, CD-Rom (2008)
7. Datta, A.K., Davidson, P.M.: Microwave and radio frequency processing. J. Food Sci. **65**, 32–41 (2000)
8. Ryynanen, S.: The electromagnetic properties of food materials: a review of the basic principles. J. Food Eng. **26**, 409–429 (1995)
9. Zhao, H., Turner, I.W.: The use of a coupled computational model for studying the microwave heating of wood. Appl. Math. Model. **24**, 183–197 (2000)
10. Yoshikawa, N.: Recent studies on fundamentals and application of microwave processing of material. In: Grundas, S. (ed.) Advances in Induction and Microwave Heating of Mineral and Organic Materials, pp. 3–26. Intech, Rijeka (2011)
11. Kowalski, S.J.: Continuous thermohydromechanical model using the theory of mixtures. In: Tsotsas, E., Mujumdar, A.S. (eds.) Modern Drying Technology: Computational Tools at Different Scales, vol. 1, pp. 125–154. Wiley-VCH, Weinheim (2007)
12. Malafronte, L., Lamberti, G., Barba, A.A., Raaholt, B., Holtz, E., Ahrné, L.: Combined convective and microwave assisted drying: experiments and modeling. J. Food Eng. **112**, 304–312 (2012)
13. Nastaj, J., Witkiewicz, K.: Experimental and simulation studies of the primary and secondary vacuum freeze drying at microwave heating. In: Grundas, S. (ed.) Advances in Induction and Microwave Heating of Mineral and Organic Materials, pp. 640–655. Intech, Rijeka (2011)
14. Lee, J., Kim, T.: Application of microwave heating to recover metallic elements from industrial waste. In: Grundas, S. (ed.) Advances in Induction and Microwave Heating of Mineral and Organic Materials, pp. 303–312. Intech, Rijeka (2011)
15. Ressing, H., Ressing, M., Durance, T.: Modeling the mechanisms of dough puffing during vacuum microwave drying using the finite element method. J. Food Eng. **82**, 498–508 (2007)
16. Wang, S., Tang, J., Johnson, J.A., Mitcham, E., Hansen, J.D., Hallman, G., Drake, S.R., Wang, Y.: Dielectric properties of fruits and insect pests as related to radio frequency and microwave treatments. Biosyst. Eng. **85**(2), 201–212 (2003)
17. Souraki, B.A., André, A., Mowlac, D.: Mathematical modeling of microwave-assisted inert medium fluidized bed drying of cylindrical carrot samples. Chem. Eng. Process. **48**, 296–305 (2009)
18. Das, S., Mukhopadhyay, A.K., Datta, S., Basu, D.: Prospects of microwave processing: an overview. Bull. Mater. Sci. **32**(1), 1–13 (2009)
19. Thostenson, E.T., Chou, T.W.: Microwave processing: fundamentals and applications. Part A Compos. **30**, 1055–1071 (1999)
20. Das, I., Kumar, G., Shah, N.G.: Microwave heating as an alternative quarantine method for disinfestation of stored food grains. Int. J. Food Sci. Article ID 926468, 13 pages (2013)
21. Feng, H., Yin, Y., Tang, J.: Microwave drying of food and agricultural materials: basics and heat and mass transfer modeling. J. Food Eng. Reviews. **4**(2), 89–106 (2012)
22. Campañone, L.A., Zaritzky, N.E.: Mathematical analysis of microwave heating process. J. Food Eng. **69**, 359–368 (2005)
23. Di, P., Chang, D.P.Y., Dwyer, H.A.: Heat and mass transfer during microwave steam treatment of contaminated soils. J. Environ. Eng. **126**(12), paper no. 22131 (2000)
24. Bingol, G., Pan, Z., Roberts, J.S., Devres, Y.O., Balaban, M.O.: Mathematical modeling of microwave-assisted convective heating and drying of grapes. Int. J. Agric. Biol. Eng. **1**(2), 46–54 (2008)
25. Pandit, R.B., Prasad, S.: Finite element analysis of microwave heating of potato-transient temperature profiles. J. Food Eng. **60**, 193–202 (2003)
26. Lin, Y.E., Anantheswaran, R.C., Puri, V.M.: Finite element analysis of microwave heating of solid foods. J. Food Eng. **25**, 85–112 (1995)
27. Jia, L.W., Islam, MdR, Mujumdar, A.S.: A simulation study on convection and microwave drying of different food products. Dry. Technol. **21**(8), 1549–1574 (2003)

28. Sanga, E.C.M., Mujumdar, A.S., Raghavan, G.S.V.: Simulation of convection-microwave drying for a shrinking material. Chem. Eng. Process. **41**, 487–499 (2002)
29. Arballo, J.R., Campañone, L.A., Mascheroni, R.H.: Modeling of microwave drying of fruits. Dry. Technol. **28**, 1178–1184 (2010)
30. Kristiawan, M., Sobolik, V., Klima, L., Allaf, K.: Effect of expansion by instantaneous controlled pressure drop on dielectric properties of fruits and vegetables. J. Food Eng. **102**, 361–368 (2011)
31. Sipahioglu, O., Barringer, S.A.: Dielectric properties of vegetables and fruits as a function of temperature, ash, and moisture content. J. Food Sci. **68**(1), 234–239 (2003)
32. Hasna, A.M.: Composite dielectric heating and drying: The computation process. In: 16th International Drying Symposium (IDS 2008), Hyderabad, India, CD-Rom (2008)
33. Shalmashi, A.: Drying of wet salicylic acid by microwave heating. In: 16th International Drying Symposium (IDS 2008), Hyderabad, India, CD-Rom (2008)
34. Monzó-Cabrera, J., Díaz-Morcillo, A., Catalá-Civera, J.M., De los Reyes, E.: Heat and mass transfer characterisation of microwave drying of leather. In: 12th International Drying Symposium (IDS'2000), Noordwijkerhout, The Netherlands, Paper No. 27 (2000)
35. Martín, M.E., Albors, A., Martínez-Navarrete, N., Chiralt, A., Fito. P.: Micro-structural changes in apple tissue subjected to combined air-microwave drying. In: 12th International Drying Symposium (IDS'2000), Noordwijkerhout, The Netherlands, Paper No. 419 (2000)
36. Raghavan, G.S.V., Li, Z., Wang, N., Gariepy, Y.: Control of microwave drying process through aroma monitoring. In: 16th International Drying Symposium (IDS 2008), Hyderabad, India, CD-Rom (2008)
37. Péré, C., Rodier, E., Louisnard, O., Schwartzentruber, J.: Investigations on microwave and vacuum drying experiments: drying of a model porous medium at laboratory scale. In: 12th International Drying Symposium (IDS'2000), Noordwijkerhout, The Netherlands, Paper No. 116 (2000)
38. Abhayawick, L., Laguerre, J.C., Duquenoy, A.: Runaway heating of onions during microwave drying. In: 12th International Drying Symposium (IDS2000), Noordwijkerhout, The Netherlands, Paper No. 158 (2000)
39. Emam-Djomeh, Z., Nazemi, Sh., Niakowsari, M., Ascari, G.: Combined effects of coating and microwave assisted hot air drying on kinetics of color changes of strawberry. In: 16th International Drying Symposium (IDS 2008), Hyderabad, India, CD-Rom (2008)
40. Zhang, M., Jiang, H., Lim, R.-X.: Recent developments in microwave-assisted drying of vegetables, fruits, and aquatic products-drying kinetics and quality considerations. Dry. Technol. **28**, 1307–1316 (2010)
41. González, A.B.B., Huelga, O.I., Bueno de las Heras, J.: Determination and analysis of the kinetics of drying for the combined convective-microwave drying of agar gels. In: 12th International Drying Symposium (IDS'2000), Noordwijkerhout, The Netherlands, Paper No. 157 (2000)
42. Askari, G.R., Tahmasbi, M., Emam-Djomeh Z.: Effect of drying method on drying curves, texture and microstructure of apple and tomato slices. In: 16th International Drying Symposium (IDS 2008), Hyderabad, India, CD-Rom (2008)
43. Torringa, H.M., Neijnens, H.P.M., Bartels, P.V.: Novel Process for the drying of sugar cubes applying microwave technology. In: 12th International Drying Symposium (IDS'2000), Noordwijkerhout, The Netherlands, Paper No. 230 (2000)
44. Ruiz, G., Martínez-Monzó, J., Barat, J.M., Chiralt, A., Fito, P.: Applying microwave in drying of orange slices. In: 12th International Drying Symposium (IDS′2000), Noordwijkerhout, The Netherlands, Paper No. 239 (2000)
45. Ahrens, F.W., Habeger, C.C.: Use of new applicator design ideas to improve uniformity of paper drying via microwave energy. In: 12th International Drying Symposium (IDS′2000), Noordwijkerhout, The Netherlands Paper No. 298 (2000)
46. Berteli, M.N., Marsaioli, Jr. A.: Evaluation of a continuous rotary air dryer assisted by microwaves in comparison with the conventional air dryer in the short cut pasta production. In:

13th International Drying Symposium (IDS'2002), Beijing, China, vols. A, B and C, Paper No. 874 (2002)
47. Beke, J., Kurják, Z., Bihercz, G.: Microwave field test of inner moisture and temperature conditions of beetroot. In: 13th International Drying Symposium (IDS'2002), with the Conventional Air Dryer in the Short Cut Pasta Production, Beijing, China, vols. A, B and C, Paper No. 891 (2002)
48. RodrÍGuez, R., De Elvira, C., Lombraña, J.I., Kamel, M.: Effect of different vacuum microwave drying strategies on moisture sorption capacity of freeze-dried products. In: 13th International Drying Symposium (IDS'2002), Beijing, China, vols. A, B and C, Paper No. 909 (2002)
49. Dong, T., Kimura, T., Yoshizaki, S.: Microwave drying of thick layer brown rice with concurrent flow ventilation. In: 13th International Drying Symposium (IDS'2002), Beijing, China, 2002, vols. A, B and C, Paper No. 917 (2002)
50. Mcloughlin, C.M., Mcminn, W.A.M., Magee, T.R.A.: Combined microwave and convective drying of pharmaceuticals. In: 13th International Drying Symposium (IDS'2002), Beijing, China, vols. A, B and C, Paper No. 1472 (2002)
51. Lescano, C., César, L., Tonin, L., Pereira, N., Marsaioli, A.: Kinetics evaluation of the microwave drying of Okara. In: 16th International Drying Symposium (IDS'2008), Hyderabad, India, CD-Rom (2008)
52. Nordin, M.F.M., Gariépy, Y., Daud, W.R.W., Raghavan, G.S.V., Talib, M.Z.M.: Microwave and hot air drying for red pitaya (hylocereus undatus). In: 16th International Drying Symposium (IDS'2008), Hyderabad, India, CD-Rom (2008)
53. Shalmashi, A.: Microwave assisted drying of wet caffeine. In: 16th International Drying Symposium (IDS'2008), Hyderabad, India, CD-Rom (2008)
54. McMinn, W.A.M., McLoughlin, C.M., Farrell, G., Magee, T.R.A.: Temperature profiles in powder beds during microwave drying. In: 14th International Drying Symposium (IDS'2004), São Paulo, Brazil, vol. B, pp. 1081–1088 (2004)
55. Wang, W., Chen, G.: Theoretical study on microwave freeze-drying of an aqueous solution of lactose with the aid of dielectric material. In: 14th International Drying Symposium (IDS'2004), São Paulo, Brazil, vol. B, pp. 1142–1149 (2004)
56. Itaya, Y., Uchiyama, S., Hatano. S., Mori, S.: Drying enhancement of clay slab by microwave heating. In: 14th International Drying Symposium (IDS 2004), São Paulo, Brazil, vol. A, pp. 193–200 (2004)
57. Andrés, A.M., Betoret, N.N., Guillermo, B., Fito, P.: Microwave-assisted air drying of cassava slices. In: 14th International Drying Symposium (IDS'2004), São Paulo, Brazil, vol. C, pp. 1599–1603 (2004)
58. .Bihercz, G., Kurjak, Z.: Analysis of the microwave and convective vegetable dewatering process as a function of drying conditions. In: 14th International Drying Symposium (IDS'2004), São Paulo, Brazil, vol. C, pp. 1652–1659 (2004)
59. Kamel, M., De Elvira, C., Lombraña, J.I., RodrÍGuez, R.: Drying kinetics and energy consumption in vacuum drying process with microwave and radiant heating. In: 13th International Drying Symposium (IDS'2002), Beijing, China, vols. A, B and C, Paper No. 882 (2002)
60. Qianqian, D., Junhong, Y., Wei, W., Mingdi, S., Jun, Z.: Effect of microwave drying technology on micro-structure and dehydration characteristic of angelica sleces. In: 16th International Drying Symposium (IDS'2008), Hyderabad, India, CD-Rom (2008)
61. Yarmand, M.S., Homayouni, A.: Microware processing of meat. In: Chandra, U. (ed.) Microwave Heating, pp. 107–134. Intech, Rijeka (2011)
62. Shaheen, M.S., El-Massry, K.F., El-Ghorab, A.H., Anjum, F.M.: Microwave applications in thermal food processing. In: The Development and Application of Microwave Heating, pp. 3–16. Intech, Rijeka (2012)
63. Saxena, V.K., Chandra, U.: Microwave synthesis: a physical concept. In: Chandra, U. (ed.) Microwave Heating, pp. 3–22. Intech, Rijeka (2011)
64. Kreith, F., Bohn, M.S.: Principle of Heat Transfer. Brooks/Cole, Pacific Grove (2001)

65. Kaviany, M.: Principle of Heat Transfer. Wiley, New York (2002)
66. Strumillo, C., Kudra, T.: Drying: principles, science and design. Gordon and Breach Science Publishers, New York (1986)
67. Ratti, C., Mujumdar, A.S.: Infrared drying. In: Mujumdar, A.S. (ed.) Handbook of Industrial Drying, pp. 423–438. CRC Press, Boca Raton (2007)
68. van't Land, C.M.: Industrial Drying Equipment: Selection and Application. Marcel Dekker Inc., New York (1991)
69. Incropera, F.P., DeWitt, D.P.: Fundamentals of Heat and Mass Transfer. Wiley, New York (2002)
70. Seyed-Yagoobi, J., Sikirica, S.J., Counts, K.M.: Heating/drying of paper sheet with gas-fired infrared emitters-pilot machine trials. In: 12th International Drying Symposium (IDS′2000), Noordwijkerhout, The Netherlands, Paper No. 319 (2000)
71. Noboa, H., Seyed-Yagoobi, J.: Drying of uncoated paper with gas-fired infrared emitters-optimum emitters' location within the paper machine drying section. In: 13th International Drying Symposium (IDS"2002), Beijing, China, vols. A, B and C, Paper No. 81 (2002)
72. Seyed-Yagoobi, J., Noboa, H.: Heating/drying of uncoated paper with gas-fired and electric infrared emitters—fundamental understanding. In: 14th International Drying Symposium (IDS'2004), São Paulo, Brazil, vol. B, pp. 1217–1224 (2004)
73. Alves-Filho, O., Eikevik, T.M.: Hybrid heat pump drying technologies for porous materials. In: 16th International Drying Symposium (IDS'2008), Hyderabad, India, CD-Rom (2008)
74. Allanic, N., Salagnac, P., Glouannec, P., Ploteau, J.P.: Experimental study of the drying and curing of water based varnishes by infrared radiation. In: 16th International Drying Symposium (IDS'2008), Hyderabad, India, CD-Rom (2008)
75. Dontigny, P., Angers, P., Supino, M.: Graphite slurry dehydration by infrared radiation under vacuum conditions. In: International Drying Symposium (IDS'92), Part A, pp. 669–678 (1992)
76. Brooks, F.A., Miller, W.: Availability of solar energy. In: Zarem, A.M., Erway, D. (eds.) Introduction to the Utilization of Solar Energy, pp. 30–58. McGraw-Hill, New York (1963)
77. McVeigh, J.C.: Sun power—an introduction to the application of solar energy. Pergamon Press, New York (1977)
78. Sabins, F.F.: Remote Sensing: Principles and Interpretation. W.H. Freeman, San Francisco (1978)
79. Michels, T.: Solar Energy Utilization. Van Nostrand Reinhold, New York (1979)
80. WHO (World Health Organization). Environmental Health Criteria 14: Ultraviolet Radiation. WHO, Geneva (1979)
81. Akinjiola, O.P., Balachandran, U.: Mass-heater supplemented greenhouse dryer for post-harvest preservation in developing countries. J. Sustain. Dev. **5**(10), 40–49 (2012)
82. Bansal, N.K., Garg, H.P.: Solar crop drying. In: Mujumdar, A.S. (ed.) Advances in Drying, Hemisphere Publishing Corporation, Washington, DC, vol. 4, pp. 279–358 (1987)
83. Imre, L.: Solar drying. In: Mujumdar, A.S. (ed.) Handbook of Industrial Drying, 3rd edn, pp. 307–361. CRC Press, Boca Raton (2007)
84. Sankat, C.K., Mujaffar, S.: Sun and solar cabinet drying of salted shark fillets. In: 14th International Drying Symposium (IDS'2004), São Paulo, Brazil, vol. C, pp. 1584–1591 (2004)
85. Weiss, W., Buchinger, J.: Solar drying. Training course within the scope of the project: establishment of a production, sales and consulting infrastructure for solar thermal plants in Zimbabwe, Arbeitsgemeinschaft ERNEUERBARE ENERGIE Institute for Sustainable Technologies: AEEIntec, Austria Development Corporation, Gleisdorf, Austria. p. 110. Retrieved from http://www.aee-intec.at/0uploads/dateien553.pdf. Accessed 12 April 2015
86. Ogheneruona, D.E., Yusuf, M.O.L.: Design and fabrication of a direct natural convection solar dryer for tapioca. Leonardo Electr. J. Pract. Technol. **10**(18), 95–104 (2011)
87. Gibriel, A.Y., El-Sahrigi, A.F., Rasmy, N.M., Heikal, Y.A., Ibrahim, H.K.: Dehydration of apricots and grapes using solar dehydration. In: 12th International Drying Symposium (IDS'2000), Noordwijkerhout, The Netherlands, Paper No. 392 (2000)

88. Sablani, S,S., Rahman, M.S., Mahgoub, O., Al-Marzouki, A.S.: Sun and solar drying of fish sardines. In: 13th International Drying Symposium (IDS'2002), Beijing, China, vols. A, B and C, Paper No. 1662 (2002)
89. Sakin, M., Mcminn, W.A.M., Magee, T.R.A.: Drying behaviour of biscuit baking in convectional and combined microwave and convectional ovens. In: 13th International Drying Symposium (IDS'2002), Beijing, China, vols. A, B and C, Paper No. 1676 (2002)
90. Darvishi, H., As, A.R., Asghari, A., Najafi, G., Gazori, H.A.: Mathematical modeling, moisture diffusion, energy consumption and efficiency of thin layer drying of potato slices. J. Food Process. Technol. **4**(3), 1–6 (2013)
91. Li, X.J., Zhang, B.-G., Li, W.J.: Microwave-vacuum drying of wood: model formulation and verification. Dry. Technol. **26**, 1382–1387 (2008)
92. Dinčov, D.D., Parrott, K.A., Pericleous, K.A.: Heat and mass transfer in two-phase porous materials under intensive microwave heating. J. Food Eng. **65**, 403–412 (2004)
93. Karaaslan, S.N., Tunçer, I.K.: Development of a drying model for combined microwave–fan-assisted convection drying of spinach. Biosyst. Eng. **100**, 44–52 (2008)
94. Kowalski, S.J., Rajewska, K., Rybicki, A.: Stresses generated during convective and microwave drying. In: 14th International Drying Symposium (IDS'2004), São Paulo, Brazil, vol. A, pp. 351–358 (2004)
95. Chen, P., Schimidt, P.S.: Mathematical modeling of dielectrically—enhanced drying. In: Turner, I., Mujumdar, A.S. (eds.) Mathematical Modeling and Numerical Techniques in Drying Technology, pp. 439–479. Marcel Dekker Inc, New York (1997)

Heat and Mass Transfer, Energy and Product Quality Aspects in Drying Processes Using Infrared Radiation

Rodolfo J. Brandão, Lidja D.M.S. Borel, Luanda G. Marques and Manoel M. Prado

Abstract The high cost of the energy consumption in conventional drying plants has encouraged the search for more economically attractive drying technologies that provide improvements in energy utilization, in order to accelerate the process and to reduce operating costs. The aim of this chapter is to highlight through experimental studies the pros and cons of using IR radiation, and its applicability for drying different types of materials. Two case studies are presented, involving the use of infrared radiation alone or combined to forced air convection in drying processes of different grains and seeds. The performance of the drying systems, in terms of drying kinetics, energy consumption and seed quality is evaluated and compared to that from conventional hot-air convection dryer.

Keywords Drying · Experimental · Infrared · Energy · Quality aspects

1 Introduction

Drying is an essential operation in the chemical, agricultural, biotechnology, food, polymer, ceramic, pharmaceutical, pulp and paper, and wood processing industries. Processing of different materials by hot air drying is still the most common and most costly and energy-consuming industrial step. It accounts for up to 15 % of

---

R.J. Brandão · L.D.M.S. Borel · L.G. Marques · M.M. Prado (✉)
Department of Chemical Engineering, Federal University of Sergipe,
Marechal Rondon Avenue, São Cristóvão, 49100-000 Sergipe, Brazil
e-mail: manoelprado@ufs.br

R.J. Brandão
e-mail: rodolfoj.brandao@hotmail.com

L.D.M.S. Borel
e-mail: dahiane.ms@hotmail.com

L.G. Marques
e-mail: luanda_gimeno@yahoo.com.br

J.M.P.Q. Delgado and A.G. Barbosa de Lima (eds.),
Drying and Energy Technologies, Advanced Structured Materials 63,
DOI 10.1007/978-3-319-19767-8_6

industrial energy usage [34]. Convective drying is a highly energy-intensive process, due to the high latent heat of vaporization and the inherent inefficiency of using hot air as the drying medium [26].

Efficient use of energy in such energy-intensive operations as drying is crucial to the reduction of net energy consumption. Current research in drying is therefore largely directed towards development of cost-effective drying processes using alternative and efficient energy sources to replace or help convection heating.

Use of microwaves energy, ultrasound energy and infrared radiation alone or combined to convective heating are recent strategies that have been investigated [6, 13, 18] to allow performing drying of different materials in a fast and efficient manner.

According to Salagnac et al. (2004), radiating technologies are particularly interesting to the optimization of thermal conditions and hence to the process intensification, since they allow immediate and significant energy input to the product to be processed, without heating the air surrounding.

The technology using microwaves yields very fast drying with good product quality, but it is expensive and the overall efficiency is greatly dependent upon the conversion efficiency of the electrical energy into heat. Infrared (IR) radiation is one of the increasingly popular, but not yet common, methods of supplying heat to the product for drying. IR dryers are now designed to utilize radiant heat as the primary source [36]. Infrared sources are inexpensive compared to dielectric or microwave sources. They are very reliable, efficient in producing infrared energy, have a long service life and require low maintenance. They are available in various shapes and can be easily fitted to the dryer geometry [11].

Unlike hot air drying, where heat is transferred by convection from the air to the material, in IR drying the material absorbs the electromagnetic wave energy directly with reduced energy loss. Heating elements like lamps or radiators produce rays/waves in the IR region that travel directly to the exposed material surface and penetrate it, being absorbed by some components of the product. The absorbed electromagnetic energy causes molecular vibration, thus generating internal heating [38]. No energy is wasted directly heating the surrounding air. Moisture inside the material is rapidly heated and vaporized. Since heating is intense, the temperature gradient in the material reduces within a short period. Infrared radiation is also advantageous as the high drying rates suppress oxidation and prevents heat-sensitive substances being lost from the product during drying. In addition, energy consumption in infrared drying process tends to be relatively lower. Other advantages over conventional heating pointed out by many authors [6, 11, 17, 27, 38] include: higher energy efficiency, shorter drying time, versatility, simplicity of equipment, fast response times, allowing easy and rapid control, easy incorporation to any existing conventional dryer, low capital cost, a reduced necessity for air flow across the product and low product deterioration.

Krishnamurthy et al. [20] and Ratti and Mujumdar [36] present comprehensive reviews of IR drying and report basic principles and recent developments and applications of this radiating technology.

According to Ratti and Mujumdar [36], in the last 15 years there was a significant increase in publications on IR drying, in which IR radiation has been reported to be successfully applied in the drying of various materials, such as: paper [40], polymers [16, 35], pharmaceutical films [21], paints and pigments[2, 32], hydrous ferrous sulphate [13], food and agricultural products [1, 17, 27].

Seeds and grains appears to be very attractive for infrared drying as it usually contains high levels of moisture, protein and lipid which have good absorption of electromagnetic energy in the infrared region [38]. Due to small size of particles, penetration capability of IR radiation tends to be very good, what can contribute for uniform heating and reduced moisture gradient. IR radiation has been usually applied to dry grains in a single or thin layer [1, 4]. In studies on deep bed drying, vibration or fluidization has been used to ensure uniform heating of particles exposed to IR radiation, to prevent overheating and thermal stress development, thus avoiding cracking [3, 9, 10]. Moreover, IR heating has also been used for simultaneous drying and disinfestation of grains [30].

Besides technical and economical advantages due drying process intensification by applying IR radiation, preserving product quality is also a prime concern in IR drying. It could result in poor quality products if not adequately applied. Improperly chosen drying conditions, such as a high intensity of IR radiation and a long time of exposure to this electromagnetic energy may cause overheating and damages to the material. Thus, it is important not only to control IR radiation intensity, but also continuously monitoring temperature and some quality attributes of the product.

Some works state that using IR radiation for drying heat-sensitive materials is only effective in combination with other drying methods [31]. As consequence, hybrid drying has also been the subject of extensive investigation [14, 19]. The application of combined electromagnetic radiation and hot air heating has been considered more efficient than radiation or hot air heating alone as it gives a synergistic effect. Combined IR radiation and hot-air drying has been reported to reduce the drying time and the energy consumption as well as to improve quality of various products, such as pollen grains [4]; kaolin [19], soybean [10], longan fruit [28], carrot slices [23]; onion [31] and mulberry leaves [41].

Spite of large number of works dealing with IR drying, few studies are devoted to provide comparison with conventional hot air drying at similar conditions. This makes hard to evaluate the true potential of IR radiation in accelerating drying process, saving energy and improving product quality over conventional convective drying.

This chapter presents two case studies related to the use of infrared radiation alone or combined to forced air convection in drying processes of different grains and seeds. The first case deals with the use of IR radiation alone to dry two different species of oilseeds and its quantitative comparison to hot air drying from the point of view of the mass transfer and energy consumption. The aim is to evaluate through experiments the potential of IR radiation in intensifying drying rate, reducing drying time and saving energy, as well as to examine its applicability for different types of materials. In the second case emphasis is put on the combined use of IR radiation and non-heated air flow as a way to control the surface temperature of heat sensitive pollen grains while keeping a good energy efficiency and enhancing product quality.

2 Case Study I: Infrared Drying of Oilseeds

Despite its high energy consumption, hot air drying is still the most common method for the agro-industry to process oilseeds. There is therefore an urgent need for improvements in energy utilization, aiming to accelerate the process and to reduce operating costs. The aim is to highlight through experiments the pros and cons of using IR radiation and its applicability for drying different types of oilseeds: sunflower seeds (*Helianthus annuus* L.) and moringa seeds (Moringa oleifera L.). The performance of the IR dryer, in terms of drying kinetics, energy consumption and seed quality is compared to that from conventional hot-air convection dryer.

2.1 Experimental Methodology

2.1.1 Materials

Sunflower seeds from variety Embrapa 122 V2000 used in this study were provided by Embrapa Experimental Station, SNT/EM/Dourados/ MS, while moringa seeds were collected in open squares (−10.965078 to −10.944600 W; −37.075373 to −37.081825 S) in the city of Aracaju, Sergipe, Brazil.

2.1.2 Material Preparation

Sunflower seeds were received from Embrapa/SNT/EM (Brazil) with a moisture content of 0.087 dry basis (d.b.), at room temperature, and free from broken grains and foreign material. Seed size distribution was firstly determined from measurements of its dimensions using a digital micrometer (Pantec, Model 13101-25) with an accuracy of 10^{-3} mm, the counting of particles number and weighting of seed mass in a analytical balance (Shimadzu, model AUY220) with an accuracy of 10^{-7} kg. Medium seeds (length between 11.5 and 13 mm) were found to represent about 60 % population in the lot. Thus they were chosen for the study, in order to ensure uniform particle size.

Medium size sunflower seeds were then subjected to a slow artificial rewetting, through contact with nearly saturated ambient air, until the harvest moisture content (about 0.25 d.b.) was attained. The rewetted samples were then tightly packed in polythene bags and kept in a room at 18–20 °C for 12 h.

Moringa seeds were removed from green pods and a manual separation of damaged seeds was performed. Then the wings were removed from each remaining unshelled seed. As they were harvested at physiological maturity state it was not necessary rewetting them. However, the particle size distribution was also determined, in order to identify the predominant seed size range to be used in the drying experiments, thus avoiding the effect of seed size variability on the studied processes. Medium size particles, with average diameter ranging from 11 to 12.5 mm, were found to represent the highest fraction in the seed samples.

The moisture content was determined by the oven method at 105°C ± 3°C for 24 h [5]. The initial moisture content of sunflower and moringa seeds used in the drying tests were (0.21 ± 0.01) and (0.70 ± 0.03) at wet basis, respectively.

2.1.3 Drying Experiments

The experimental set-up developed to study the single layer IR drying of seeds using infrared radiation is shown schematically in Fig. 1. It consists of a drying chamber equipped with a 250 W incandescent lamp. The drying chamber of 200 × 200 × 300 mm was made from wood sheet of 10 mm thickness. The chamber walls were isolated and covered with an aluminum foil. The air flow is parallel to the sample. It is funneled through a rectangular duct. The temperature of the IR source was regulated using a temperature controller, while the air velocity was fixed by regulating the fan revolution using a DC adaptor. IR source was located at a distance of 17 cm of the sample. The dryer allows for drying with different velocities of the non-heated air flow as well as with different radiation intensities by changing temperature of the IR source and the distance from the sample.

Although the dryer is equipped with a forced air convection system, in this case study we examined the application of IR radiation alone for drying the seeds.

For sunflower seeds drying tests were conducted with IR source temperatures of 65, 80 and 93 °C, under which the particles attained temperatures of 40, 50 and 60 °C, respectively. As moringa seeds contain higher moisture content and hence a higher absorptivity the emitter temperatures were adequately reduced to 51; 61 and 72 °C, for the material reached temperatures similar to those of sunflower seeds.

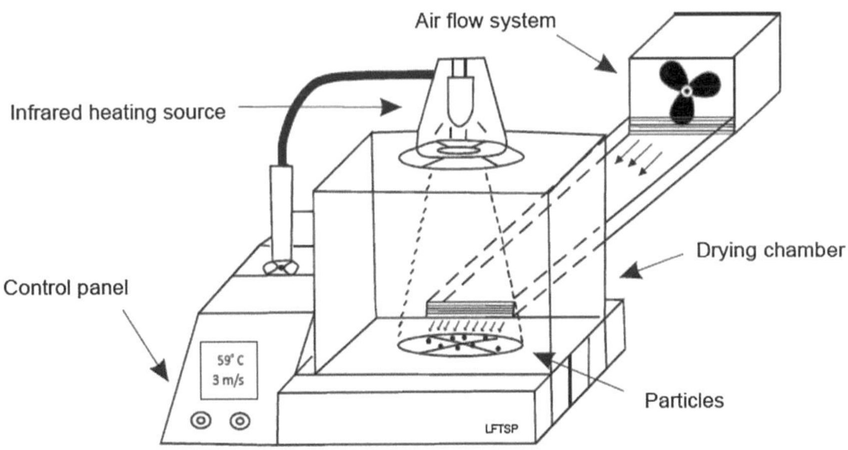

Fig. 1 Schematic diagram of the IR dryer

Conventional drying of oilseeds was also carried out for direct comparison to the IR drying characteristics. Experiments were performed in a forced air convection oven at an air velocity of 1 m/s and air temperatures of 40, 50 and 60 °C, which correspond to the final temperatures attained by the seeds in the IR drying.

To determine the drying kinetics curves the pollen samples were removed from the equipment at regular intervals along the total drying time and their weight measured in a digital balance (Sartorius, CP224S, Germany) with 10^{-4} g accuracy. All weighting processes were completed in 15 s during the drying process. Moisture content at a given time was calculated by the ratio between the mass of water retained in the material and the dry solid mass, which was determined at the end of each experiment by the oven method at (105 ± 3) °C for 24 h. The surface temperature of the particles was measured at different drying times by using an optical pyrometer.

The shrinkage of oilseeds during single layer drying was quantified from the changes in volume and surface area of individual particles, which were calculated from the dimensions taken by the digital micrometer. Reduced dimensional changes in both volume (V_p/V_{p0}) and area (A_p/A_{p0}) with respect to the moisture content constituted the so-called shrinkage curves. It must be pointed out that the shrinkage analysis, based on volume and surface area changes from the geometric measurements with the aid of a digital micrometer, was sufficiently accurate since seeds keep its original shape throughout the whole drying process.

2.1.4 Analysis of Drying Kinetics

The mass transfer in both the drying methods was analyzed and compared in terms of effective diffusivity, which was determined at each temperature by applying diffusion model to describe the drying kinetics under the consideration of both negligible and significant shrinkage.

Non-steady mass transfer equation based on Fick's second law for the diffusion of water during drying is expressed as:

$$\frac{\partial(\rho_d X)}{\partial t} = \nabla \cdot \left[D_{eff} \nabla(\rho_d X)\right] \tag{1}$$

where D_{eff} is the effective moisture diffusivity, ρ_d is the local concentration of dry solid (kg dry solid per volume of the moist material) that varies with moisture content because of shrinkage, X is the local moisture content (dry basis) and t is the drying time.

Shrinkage of sunflower seeds was found to be below 10 %. Thus, assuming one-dimensional moisture transfer with negligible shrinkage and considering that the process is isothermal, initial moisture distribution is uniform, effective water diffusivity into material is constant, and equilibrium conditions at the particle surface prevail, the analytical solution for rectangular geometry, if only the first term is taken into account, can be approximated to the form [8]:

$$XR = \frac{\overline{X} - X_e}{X_0 - X_e} = \frac{8}{\pi^2}\exp\left(-\frac{\pi^2 D_{eff}}{L_0^2}t\right) \tag{2}$$

where $\overline{X}$ is the average moisture content at a given time, X_0 and X_e are the initial and equilibrium moisture content of the seeds, respectively, L_0 is the seed thickness and D_{eff} is the effective moisture diffusivity without considering the particle shrinkage.

On the other hand, shrinkage of moringa seeds was not negligible during drying, irrespective of the applied method. Moringa seeds shrunk more than 50 % of their original dimensions, depending upon the drying conditions. In order to include the shrinkage effects into diffusion model, substituting the density of dry solid ($\rho_d = m_d/V_p$), Eq. (3) for constant mass of dry solid could be expressed as (Barbosa Neto et al. [1, 15]:

$$\frac{\partial(X/V_p)}{\partial t} = \nabla \cdot \left[D_{eff}\nabla(X/V_p)\right] \tag{3}$$

where V_p is the particle volume. Substituting $Y = X/V_p$, the following equation is obtained:

$$\frac{\partial Y}{\partial t} = \nabla \cdot \left(D_{eff}^{*}\nabla Y\right) \tag{4}$$

For moisture transfer only at the radial direction, in conjunction with the following initial and boundary conditions

$$t = 0, Y = X_0/V_0 \tag{5}$$

$$t > 0, r = 0, \frac{\partial Y}{\partial r} = 0 \tag{6}$$

$$t > 0, r = R, Y = X_{eq}/V_{eq} \tag{7}$$

the analytical solution for spherical geometry is:

$$YR = \frac{\overline{Y} - Y_e}{Y_0 - Y_e} = \frac{X/V_p - X_e/V_{pe}}{X_0/V_{p0} - X_e/V_{pe}} = \frac{6}{\pi^2}\exp\left(-\frac{\pi^2 D_{eff}^{*}}{R^2}t\right) \tag{8}$$

where V_p is the sample volume at a given time, V_{p0} is the initial volume and V_{pe} is the sample volume on attaining the equilibrium moisture, D_{eff}^{*} is the effective moisture diffusivity into the shrinking particle and R is the time averaged radius during drying.

Equations (2) and (8) are valid only for describing the falling rate period when the drying process is controlled by internal moisture diffusion and moisture content is below the critical value. Therefore, the water diffusivities into oilseeds undergoing both IR and convective drying were estimated from these equations by setting the initial moisture content to the critical value X_{cr} and by setting the drying time to zero when the mean moisture content of the sample reaches that critical moisture content. Experimental data of XR and YR were then fitted to the Eqs. (2) and (8), respectively. D_{eff} and D^*_{eff}-values were estimated by applying a non-linear regression procedure using STATISTICA® software. The estimation method was based on the well-established Levenberg-Marquardt algorithm [12].

The statistical criterion used to evaluate the goodness-of-fit of Eqs. (2) and (8) was the coefficient of determination (R^2). The dependency of effective diffusivity on temperature was investigated by plotting the obtained values against product temperature. The trends were visually inspected and appropriate model was fitted to the data.

2.1.5 Analysis of Specific Energy Consumption

The performance of each dryer with respect to the energy consumption during the process was also evaluated. Analysis was based on the specific energy consumption (SEC), which was defined as the energy required for removing unity of mass of water existing in the sunflower seeds from the initial moisture content of 0.25 ± 0.02 w.b. to a final moisture content 0.08 d.b. It was expressed as MJ/kg of moisture removed.

2.2 *Results*

2.2.1 Drying Kinetics

The mean values of the effective moisture diffusivity into sunflower and moringa seeds subjected to convective and infrared drying at the same corresponding final product temperatures are presented in Fig. 2.

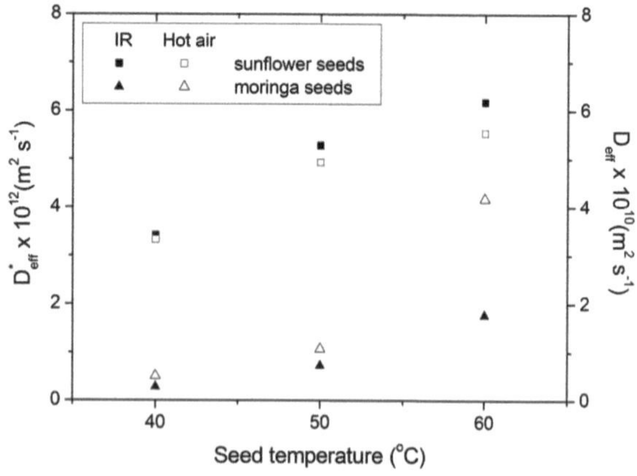

Fig. 2 Effective moisture diffusivities as a function of temperature for sunflower and moringa seeds dried by IR and hot air drying methods

The values of D_{eff} for sunflower seeds dried in hot-air tray dryer at temperatures of 40–60 °C, varied between 3.33×10^{-10} and 5.54×10^{-10} $m^2\ s^{-1}$, while those for IR-dried seeds were in the range of 3.42×10^{-10}–6.18×10^{-10} $m^2\ s^{-1}$ at the corresponding product temperatures. The R^2-values ranged from 0.973 to 0.998. These D_{eff}-values are within the range of 10^{-9}–10^{-11} $m^2\ s^{-1}$ reported in the literature for several agricultural products (Perea-Flores et al. [33].

A comparative analysis of the drying methods in terms of D_{eff}—values presented in Fig. 2 reveals that the use of IR radiation for drying sunflower seeds is only more effective than hot air convection when operated at heating source temperatures higher than 80 °C, which corresponds to material temperature higher than 50 °C. At 50 and 60 °C, the effective diffusion coefficient into infrared-dried seeds was 7 and 11 % higher than that into air-dried seeds.

It can also be noted from Fig. 2 that the application of IR radiation was not able to intensify mass transfer in drying of moringa seeds. Over the whole investigated conditions the effective moisture diffusivity values in hot air drying were higher than those obtained from IR drying. The effective diffusivities into air-dried seeds were in the range of 0.5×10^{-12}–6.18×10^{-12} $m^2\ s^{-1}$, while into IR-dried seeds they ranged from 0.28×10^{-12} to 1.77×10^{-12} $m^2\ s^{-1}$. This result can be better understood by analyzing and comparing the temporal evolutions of seed surface temperature and drying rates during tests using solely convective heating with those obtained during tests using only infrared heating, as presented in Fig. 3 and 4, respectively.

As revealed by Fig. 3, the surface temperature in moringa seeds subjected to convective drying at air temperatures of 40 and 60 °C climbs dramatically within 6–15 min after beginning the process. This step corresponds to the warming-up period, during which the drying rate also increases (Fig. 4). After this initial heating period, surface temperature of seeds remains stable at 31 and 39.5 °C during 24 and 15 min for air temperatures of 40 and 60 °C, respectively. During these times, the surface temperature is equal to the wet bulb temperature, which characterizes the

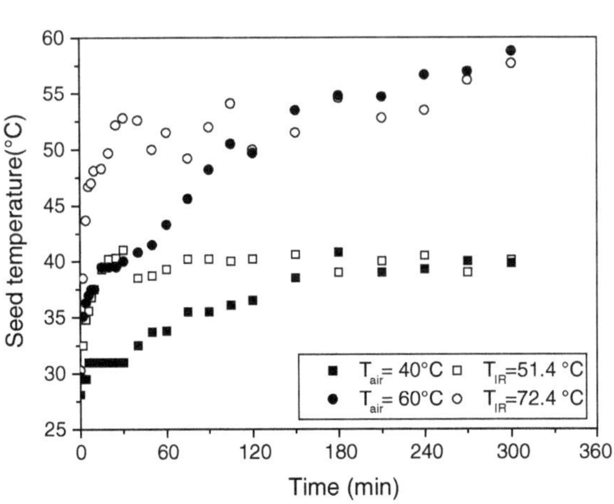

Fig. 3 Surface seed temperature as a function of time during convective and IR drying of moringa seeds

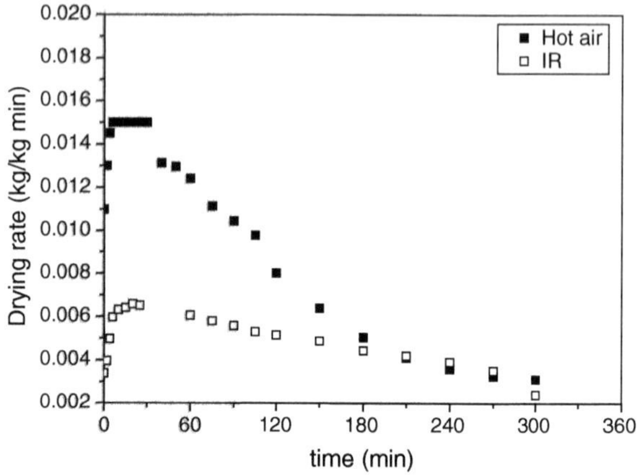

Fig. 4 Drying rate as a function of time during convective and IR drying of moringa seeds, at similar final product temperature of 40 °C

constant-rate period, in which migration of internal moisture compensates the evaporation rate from the particle surface, maintaining it saturated with a moisture film. As a consequence, all energy is used as latent heat to vaporize the surface moisture and drying occurs under a constant saturation temperature, as evidenced by the temperature plateaus in Fig. 3. At the end of the constant-rate period the aforementioned moisture compensation ceases, dried regions on the surface appear and part of the energy is used in the form of sensible heat, so that solid temperature increases, while the drying rate decreases, a typical behavior of the falling-rate period, which is mainly controlled by moisture diffusion within the particles. This is a complex mechanism involving water in both liquid and vapor states, which is commonly characterized by a so-called effective diffusivity [8].

Compared with the hot air drying, the use of IR radiation allows a more rapid rise of the surface temperature (Fig. 3), which significantly increases the saturating vapor pressure, thus activating moisture evaporation. This temperature evolution in IR-dried seeds highlights a very rapid superficial drying.

As soon as the surface dries up the internal resistance to moisture transport becomes greater than the external resistance to water vapor removal from the seed surface. In terms of the drying rate evolution (Fig. 4), this means that the drying rate drops as soon the maximum value is attained at the end of the heating period resulting in an abbreviation or even suppression of the constant-rate period.

The temperature of seeds exposed to IR energy source at 51 °C reached 40 °C just after 20 min, while the same final temperature was attained by air-dried seeds after 180 min, that is, in nine times longer time than in the IR drying, thus revealing a more gradual convective heating.

Although driving force for mass transfer depends on temperature, the extremely intensive heating of the drying surface by using IR radiating energy did not enhance the drying rate. As soon the seed surface dries up, it becomes more susceptible to over-drying and consequently to thermal damages.

The low moisture content of the external surface may induce a rubber–glass transition and the formation of a porous outer rigid crust or shell [22] that represents an extra resistance to mass transfer. This phenomenon called case hardening results in slowing down the drying of the material, explaining because the moisture removal rate of IR-dried seeds is very significantly lower than that from air-dried seeds.

It should be pointed out that moringa and sunflower seeds have different radiation properties what helps to explain the distinct results obtained when applying IR energy for drying these oilseeds. The former has initial higher moisture content and a higher particle size, thus presenting a higher absorptivity and a lower transmissivity. This may have contributed to intense surface heating, which led to case hardening.

In order to avoid overheating of the surface and to allow the radiation penetrates into the material, thus making the IR drying feasible not only for moringa but also for any seed specie, our research group has directed works to investigate appropriate intermittence periods of IR heat input as well as the effective combination of IR radiation and non-heated air flow.

2.2.2 Specific Energy Consumption of the Drying Systems

Tables 1 and 2 list the specific energy consumption (SEC) of the infrared and convective drying systems, respectively, applied for sunflower seeds at different operating conditions. It is seen that the specific energy consumption was in the range of 477.0-382.7 MJ kg^{-1} evaporated water for the hot-air dryer, and in the range of 165.-5-147.6 MJ kg^{-1} evaporated water for the IR dryer. In both methods, drying temperature was the main factor affecting energy consumption. Increased temperature decreased drying time, thus reducing the amount of energy consumed.

The use of IR radiation resulted in a reduction in the drying time by 30 % in comparison to convective drying, at the same corresponding temperature

Table 1 Specific energy consumption in convective drying of sunflower seeds at different temperatures

T_{air} (°C)	T_s (°C)	Drying time (min)	Total energy (kJ)	SEC (MJ/kg)
40	40	75	880.5	477.0
50	50	60	897.9	450.8
60	60	50	769.3	382.7

Table 2 Specific energy consumption in IR drying of sunflower seeds at different temperatures

T_{IR} (°C)	T_s (°C)	Drying time (min)	Total energy (kJ)	SEC (MJ/kg)
65	40	75	345.3	165.5
80	50	45	285.6	149.3
93	60	35	291.8	147.6

conditions. The IR drying system was found to offer from 60 to 65 % saving in energy consumption, showing considerable potential for improving energy efficiency and operating cost in sunflower seeds drying industry. In the work of Motevali et al. [25], IR drying of pomegranate arils led to reductions in energy consumption, between 68 and 75 %, as compared to hot air convection drying.

Technical and economical advantages of the drying process intensification by applying IR radiation were accompanied by the quality of sunflower seeds, in terms of seed germination and fatty acid composition of seed oil, which was acceptable and comparable to that from the traditional hot air dryer, as detailed in Santos [39].

3 Case Study II: Heat and Mass Transfer, Energy and Product Quality Aspects in Combined Infrared-Convective Drying of Bee-Pollen

The focus of this second case study is to evaluate the application of a hybrid dryer combining the IR radiation with non-heated air flow to prevent over-drying and to improve quality of dehydrated bee-pollen. The effects of the IR radiation intensity and air velocity on the temporal evolutions of grain temperature and moisture content, specific energy consumption and product color are then presented.

3.1 Experimental Methodology

3.1.1 Materials

Honeybee-collected pollen from apiaries located in Neópolis (State of Sergipe) was used in this study. Pollen grains *in nature* were obtained directly from the producers with a moisture content ranging between 0.18 and 0.22 wet basis (w.b.). Immediately after receiving the samples, the particle size distributions were determined by sieve analysis. Grains with medium sieve diameter of 2.03 mm were found to represent about 60 wt%. Thus, they were chosen to be used in the drying tests, in order to avoid the effect of particle size variability on heat and mass transfer during drying. Pollen samples were then stored in a commercial freezer until use.

3.1.2 Drying Experiments

The experiments in the hybrid dryer presented in Fig. 1 were conducted with pollen grains arranged in a thin layer and exposed to three IR radiation intensity levels (150, 300 and 450 W/m^2) and three air velocity levels (1.0, 2.0 and 3.0 m/s) at 25 °C.

3.1.3 Analysis of Mass Transfer

The mass transfer in hybrid drying of bee-pollen was analyzed in terms of effective diffusivity, which was determined at each set of operating conditions by applying diffusion model in spherical geometry, without considering shrinkage of the grains, to describe the drying kinetics.

3.1.4 Specific Energy Consumption

To evaluate the performance of IR dryer, the energy consumption during the process was determined. It was the sum of energy consumed by IR lamps and fan used for moving the air. Analysis was based on the specific energy consumption (SEC), which was defined as the energy required for removing unit mass of water existing in the material from the initial moisture content of 0.20 ± 0.02 w.b. to a final moisture content 0.05 w.b. It was expressed as MJ/kg of moisture removed.

3.1.5 Color Measurements

The color of the fresh and dehydrated pollen was assessed by using a MiniScan Ez Chroma Meter (Hunterlab) with a CIE standard illuminant C to determine CIE color space co-ordinates, L*, a* and b* values. All measurements were made with a light source D65 area and the observer at 10º. The standard conditions were established for the emission and detection of the light beam.

In the CIEL*a*b* color space, the L* value represents the degree of brightness, how dark/light the sample is, varying from 0 (black) to 100 (white). a* value gives the degree of greenness (0 to −60) or redness(0 to + 60), while the b* value indicates the degree of blueness (0 to −60) or yellowness (0 to +60).

The browning index (BI) was determined using the following equation [29]:

$$IE = \frac{100(x - 0.031)}{0.17} \tag{9}$$

where,

$$x = \frac{a^* + 1.75L^*}{5.645L^* + a^* - 3.012b^*} \tag{10}$$

The browning index measures the magnitude of the brow color, indicating the occurrence of enzymatic and non-enzymatic browning reactions, which produce brown pigments in the material. A larger value denotes greater browning of the sample.

3.1.6 Effects of Process Parameters

The effects of process parameters on the specific energy consumption (SEC), moisture diffusivity (D_{eff}) and color parameters of pollen grains was assessed by response surface methodology, as described in detail by Montgomery [24]. The response of dependent variables (SEC and D_{eff}) was correlated to the independent variables (IR heat flux and air velocity) by second order polynomial equations of the type:

$$Y = \alpha_0 + \alpha_1 X_1 + \alpha_2 X_2 + \alpha_3 X_1 X_2 + \alpha_4 X_1^2 + \alpha_5 X_2^2 \tag{11}$$

where Y is the response (the specific energy consumption or the moisture diffusivity), X_1 is the independent variable for the IR heat flux, X_2 is the independent variable for the air velocity and α_0, α_1, α_2,...are the regression coefficients.

3.2 *Results*

3.2.1 Drying Kinetics

Typical results of moisture content and grain temperature versus time for bee pollen exposed to IR radiation intensity of 450 W/m^2 and different non-heated air velocities are presented in Figs. 5 and 6, respectively.

The use of IR radiation without to be assisted by non-heated air flow produced a steeper drying curve (Fig. 6) resulting in a fast moisture removal. This can be attributed to a significant increase in particle temperature, as shown in Fig. 5, resulting into an increase in the water vapor pressure inside the product and thus in higher drying rates [31].

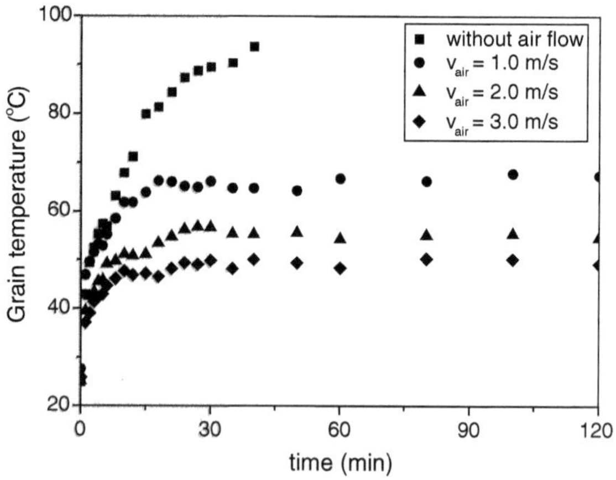

Fig. 5 Grain temperature as a function of time during hybrid drying of bee-pollen

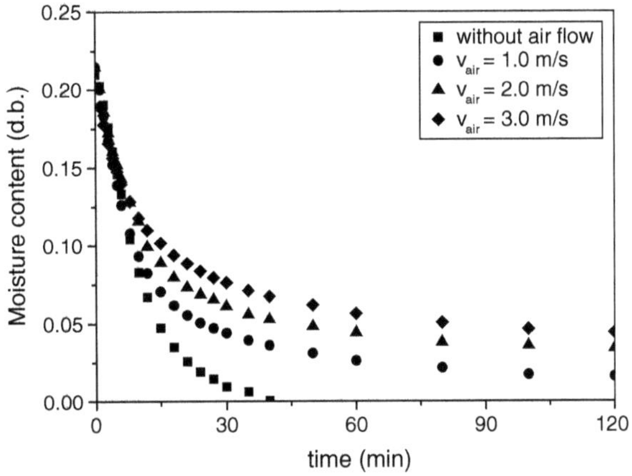

Fig. 6 Moisture content as a function of time during hybrid drying of bee-pollen

It can be seen from Fig. 6 that the moisture removal was found to be lower with higher air velocities. This result provides evidences on the cooling effect of air flow as reported by Motevali et al. [25]. The evolution of particle temperature (Fig. 5) during drying combining IR energy with streams of air at 1, 2 and 3 m/s corroborate this effect. The increase in air velocity reduced the temperature at the surface of particles, decreasing the water vapor pressure and thus decreasing the driving force for mass transfer. However, the simultaneous use of non-heated air forced convection on pollen grains exposed to IR radiation provides an added benefit for heat-sensitive substances present in bee-pollen. As IR energy is used to heat the material inside and moves moisture to the surface, the stream of non-heated air is used to remove it from the surface, keeping the particle temperature lower, thus preventing overheating and thermal damage to the material.

3.2.2 Effects of Process Parameters on Effective Diffusivity, Specific Energy Consumption and Grain Browning

The analysis of drying kinetics in terms of the particle temperature and moisture content evolutions indicates that hybrid drying of pollen took place predominantly in the falling rate period, being therefore controlled by the moisture diffusion inside the solid.

From the response surface presented in Fig. 7 for effective diffusivity it can be noted that IR heat flux and air velocity are important factors affecting the mass transfer parameter in hybrid drying. Increased radiation intensity and decreased air velocity resulted in higher D_{eff}-values, thus intensifying the moisture transfer rates.

The reduction in D_{eff}-values with an increase in air velocity, for any particular IR radiation intensity, is related to cooling ability of air flow on the particles surface, as previously discussed.

Fig. 7 Effective moisture diffusivity in pollen grains as a function of IR heat flux and air velocity

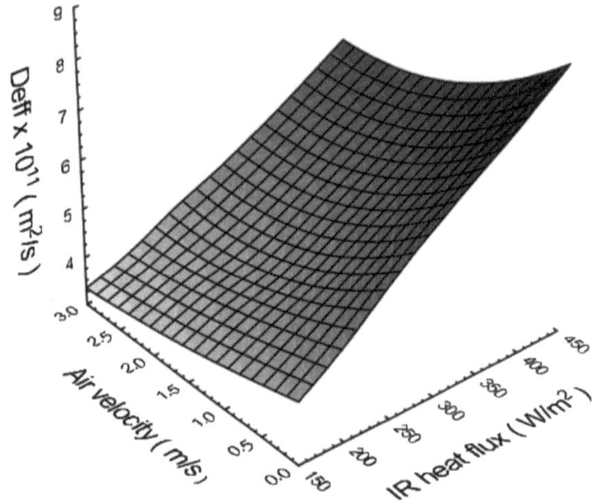

Using multivariate regression analysis, a relationship ($R^2 = 0.95$) was established between effective diffusivity, IR heat flux and air velocity:

$$D_{eff} = 2.96 \cdot 10^{-11} + 1.08 \cdot 10^{-13} IR - 5.69 \cdot 10^{-12} v_{air} \quad (12)$$

in which D_{eff} is the effective diffusivity (m^2/s), IR is the radiation intensity (W/m^2) and v_{air} is the air velocity (m/s).

The results corresponding to specific energy consumption involved in the hybrid drying of pollen, under different operating conditions, are presented in Fig. 8. The

Fig. 8 Specific energy consumption as a function of IR heat flux and air velocity

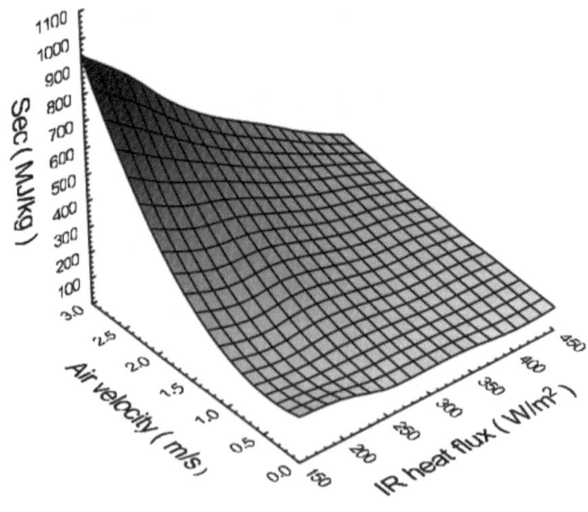

response surface shows as expected that at all air velocities increased radiation intensity accelerated the drying process, resulting in decreased specific energy consumption (SEC). When the infrared heat flux increased from 150 to 450 W/m^2 the reduction in SEC was about 78 and 55 % for air velocities of 3 and 1 m/s, respectively. For any particular radiation intensity, SEC was found to increase with the combined use of forced air convection at velocities ranging from 1 to 3 m/s. This increase was about 143 % at the maximum radiation intensity of 450 W/m^2 and 557 % at the minimum radiation intensity of 150 W/m^2. The increased energy consumption with increasing air velocity can be attributed to higher pollen surface cooling at higher air velocities, which decreases moisture removal rate, thus increasing drying time [25]. Minimum value of SEC (126.5 MJ/kg) was observed applying IR radiation alone at intensity of 450 W/m^2, while the maximum value of 971.7 MJ/kg was observed at 150 W/m^2 and 3.0 m/s.

Using multivariate regression analysis, a relationship ($R^2 = 0.93$) was established between specific energy consumption (SEC), radiation intensity (IR) and air velocity (v_{air}):

$$SEC = 194.1 - 0.691IR + 199.5v_{air} - 0.712IR \cdot v_{air} + 52.1v_{air}^2 \qquad (13)$$

in which SEC is the specific energy consumption (MJ/kg), IR is the radiation intensity (W/m^2) and v_{air} is the air velocity (m/s). Based on statistical analysis the effects of the first order terms of all process variables including their interaction as well as the second order term of air velocity were found to be significant. In addition, the magnitude of the coefficients in Eq. (12) indicated that air velocity plays a paramount influence on the specific energy consumption.

In Table 3 are listed the values of the browning index (BI) for both pollen *in nature* and dehydrated at different operating conditions. Drying pollen using IR radiation alone led to the main color changes, with increased IR heat flux increasing the browning index.

Table 3 Browning index for fresh and dehydrated pollen

Sample	IR (W/m^2)	v_{air} (m/s)	BI (-)
Fresh pollen	–	–	250 ± 2
Dehydrated pollen	150	–	256 ± 1
	150	1.0	249 ± 1
	150	2.0	251 ± 2
	150	3.0	252 ± 1
	300	–	260 ± 2
	300	1.0	247 ± 2
	300	2.0	250 ± 1
	300	3.0	249 ± 1
	450	–	268 ± 3
	450	1.0	257 ± 3
	450	2.0	259 ± 2
	450	3.0	251 ± 1

Grains exposed to IR radiation alone attained the higher values of surface temperatures (Fig. 5), which is an important factor for explaining the changes in color of pollen grains. These color changes during IR drying are mainly consequence from non-enzymatic browning, as bee pollen is rich in sugar, amino acids and carotenoids [7]. The rates of the browning reactions increase with increasing product temperature, which occurs in higher magnitude during drying when the IR radiation is not assisted by forced air convection.

In hybrid drying, as shown in Table 3, for each IR heat flux applied, increased air velocity decreases browning indices. At 150 and 300 W/m^2, an air velocity of 1 m/s was enough to minimize color changes, while at 450 W/m^2 an air velocity of 3 m/s was required to ensure an adequate cooling of grains surface, thus prevented color degradation.

4 Final Remarks

From the examination of results obtained in the case studies the following conclusion can be drawn.

Depending on the radiation properties and physical characteristics of particles as well as on structural changes that tend to occur under moisture removal applying IR radiation may or not intensify drying processes of grains and seeds.

IR drying of moringa seeds presented a performance level below that of conventional drying. Applying IR radiation to dry high moisture seeds like moringa seeds abbreviated or even suppressed the constant-rate period. Intensive superficial drying resulted in case-hardening phenomenon, which slowed down the process.

On the other hand, it was demonstrated that drying of sunflower seeds by using IR radiation is technologically feasible, with superior performance in comparison with hot air drying, yielding drying rate intensification, energy saving and product quality preservation.

Although the use of IR radiation alone accelerates drying of pollen, resulting in a significant energy saving, it causes significant browning of the dehydrated product, because of the intensive heating. It can then be concluded that using IR radiation alone is not recommended for drying pollen. However, its combination with non-heated air flow in a hybrid drying appears as a promising alternative to obtain high quality dehydrated product at low drying time and energy cost. By cooling grain surface, the use of non-heated air flow combined to IR heating helps to prevent color degradation in pollen grains, assuring maintenance of pollen quality throughout and at end of processing.

There is a large field to be explored in drying technologies making use of IR radiation. Many efforts still need to be devoted to a greater comprehension of the best strategies of applying this radiant energy, aiming to achieve significant improvements either in the product quality or in the process time/energy consumption.

References

1. Barbosa Neto, A.M., Marques, L.G., Prado, M.M.: Mass transfer in infrared drying of gel-coated seeds. Adv. Chem. Eng. Sci. **4**, 39–48 (2014)
2. Blanc, D., Laurent, P.J., Gerard, F., Andrieu, J.: Experimental infrared drying study of a model water-based epoxy-amine painting coated on iron support. Drying Technol. **15**(6–8), 1787–1799 (1997)
3. Borel, L. D. M. S.: Application of infrared heating assisted fluidized bed dryer for processing bee pólen. MSc Dissertation, pp. 149 (2014) (In Portuguese)
4. Brandão, R.J., Santos, L.D.M., Marques, L.G., Prado, M.M.: Heat and mass transfer and energy aspects in combined infrared-convective drying of bee-pollen. Defect Diffus. Forum **364**, 9–17 (2015)
5. Brazil: Regras para análise de sementes. (1st ed.). Brasília: Ministério da Agricultura, Pecuária e Abastecimento (2009)
6. Chua, K.J., Chou, S.K., Mujumdar, A.S., Ho, J.C., Hon, C.K.: Radiant-convective drying of osmotic treated agro-products: effect on drying kinetics and product quality. Food Control **15**, 145–158 (2004)
7. Chunthaworn, S., Achariyaviriya, S., Namsanguan, S.: Color kinetics of longan flesh drying at high temperature. Procedia Eng **32**, 104–111 (2012)
8. Crank, J.: The mathematics of diffusion, 2nd edn. Clarendon Press, London (1975)
9. Das, I., Das, S.K., Bal, S.: Drying kinetics of high moisture paddy undergoing vibration-assisted infrared (IR) drying. J. Food Eng. **95**(1), 166–171 (2009)
10. Dondee, S., Meeso, N., Soponronnarit, S., Siriamornpun, S.: Reducing cracking and breakage of soybean grains under combined near-infrared radiation and fluidized-bed drying. J. Food Eng. **104**, 6–13 (2011)
11. Dostie, M., Seguin, J.N., Maure, D., Tonthat, Q.A., Chatingy, R.: Preliminary measurements on the drying of thick porous materials by combinations of intermittent infrared and continuous convection heating. In: Mujumdar, A.S., Roques, M.A. (eds.) Drying'89. Hemisphere Press, New York (1989)
12. Fletcher, R.: Practical methods of optimization, 2nd edn. Wiley, London (2001)
13. Glouannec, P., Salagnac, P., Guézenoc, H., Allanic, N.: Experimental study of infrared-convective drying of hydrous ferrous sulphate. Powder Technol. **187**(3), 280–288 (2008)
14. Hammouda, I., Mihoubi, D.: Comparative numerical study of kaolin clay with three drying methods: Convective, convective–microwave and convective infrared modes. Energy Convers. Manag. **87**, 832–839 (2014)
15. Hashemi, G., Mowla, D., Kazemeini, M.: Moisture diffusivity and shrinkage of broad beans during bulk drying in an inert medium fluidized bed dryer assisted by dielectric heating. J. Food Eng. **92**(3), 331–338 (2009)
16. Kajtna, J., Sebenik, U., Krajnc, M., Golob, J.: IR drying of water-based acrylic PSA adhesives. Drying Technol. **26**(3), 323–333 (2008)
17. Kocabiyik, H., Tezer, D.: Drying of carrot slices using infrared radiation. Int. J. Food Sci. Technol. **44**(5), 953–959 (2009)
18. Kowalski, S.J., Rajewska, K.: Convective drying enhanced with microwave and infrared radiation. Drying Technol. **27**(7–8), 878–887 (2009)
19. Kowalski, S.J., Rajewska, K.: Effectiveness of hybrid drying. Chem. Eng. Process. **48**(8), 1302–1309 (2009)
20. Krishnamurthy, K., Khurana, H.K., Jun, S., Irudayaraj, J., Demirci, A.: Infrared heating in food processing: an overview. Compr. Rev Food Sci. Food Saf. **7**, 1–13 (2008)
21. Le Person, S., Puiggali, J.R., Baron, M., Roques, M.: Near infrared drying of pharmaceutical thin films: experimental analysis of internal mass transport. Chem. Eng. Process. **37**(3), 257–263 (1998)

22. Mayor, L., Sereno, A.M.: Modelling shrinkage during convective drying of food materials: a review. J. Food Eng. **61**, 373–386 (2004)
23. Mihoubi, D., Timoumi, S., Zagrouba, F.: Modelling of convective drying of carrot slices with IR heat source. Chem. Eng. Process. **48**, 808–815 (2009)
24. Montgomery, D.C.: Design and Analysis of Experiments. Wiley, USA (1997)
25. Motevali, A., Minaei, S., Khoshtagaza, M.H.: Evaluation of energy consumption in different drying methods. Energy Convers. Manag. **52**, 1192–1199 (2011)
26. Mujumdar, A.S.: Principles, classification and selection of dryers. In: Mujumdar, A. (ed.) Handbook of Industrial Drying, 4th edn, pp. 3–30. CRC Press, Florida (2015)
27. Niamnuy, C., Nachaisin, M., Poomsa, N., Devahastin, S.: Kinetics modeling of drying and conversion/degradation of isoflavones during infrared drying of soybean. Food Chem. **133**, 946–952 (2012)
28. Nuthong, P., Achariyaviriya, A., Namsanguan, K., Achariyaviriya, S.: Kinetics and modeling of whole longan with combined infrared and hot air. J. Food Eng. **102**(3), 233–239 (2011)
29. Palou, E., López-Malo, A., Barbosa-Cánovas, G.V., Welti-Chanes, J., Swanson, B.G.: Polyphenoloxidase activity and color of blanched and high hydrostatic pressure treated banana puree. J. Food Sci. **64**, 42–45 (1999)
30. Pan, Z., Khir, R., Godfrey, L.D., Lewis, R., Thompson, J.R., Salim, A.: Feasibility of simultaneous rough rice drying and disinfestations by infrared radiation heating and rice milling quality. J. Food Eng. **84**, 469–479 (2008)
31. Pathare, P.B., Sharma, G.P.: Effective Moisture Diffusivity of Onion Slices undergoing Infrared Convective Drying. Biosyst. Eng. **93**(3), 285–291 (2006)
32. Pawar, S.B., Kumar, P.S.R., Mujumdar, A.S., Thorat, B.N.: Infrared-convective drying of organic pigments. Drying Technol. **26**(3), 315–322 (2008)
33. Perea-Flores, M.J., Garibay-Febles, V., Chanona-Perez, J.J., Calderon-Dominguez, G., Mendez-Mendez, J.V., Palacios-Gonzalez, E., Gutierrez-Lopez, G.F.: Mathematical modelling of castor oil seeds (Ricinus communis) drying kinetics in fluidized bed at high temperatures. Ind. Crops Prod. **38**, 64–71 (2012)
34. Perussello, C.A., Kumar, C., Castilhos, F., Karim, M.A.: Heat and mass transfer modeling of the osmo-convective drying of yacon roots (Smallanthus sonchifolius). Appl. Therm. Eng. **63**, 23–32 (2014)
35. Putranto, A., Chen, X.D., Webley, P.A.: Infrared and convective drying of thin layer of polyvinyl alcohol (PVA)/glycerol/water mixture—The reaction engineering approach (REA). Chem. Eng. Process. **49**, 348–357 (2010)
36. Ratti, C., Mujumdar, A.S.: Infrared Drying. In: Mujumdar, A.S. (ed.) Handbook of industrial drying, 4th edn, pp. 405–420. CRC Press, Florida (2015)
37. Salagnac, P., Glouannec, P., Le Charpentier, D.: Numerical modeling of heat and mass transfer in porous medium during combined hot air, infrared and microwaves drying. Int. J. Heat Mass Transf. **47**, 4479–4489 (2004)
38. Sandu, C.: Infrared radiative drying in food engineering: a process analysis. Biotechnol. Prog. **2**(3), 109–119 (1986)
39. Santos, C. J. R.: Drying of sunflowerseeds using infrared radiation and forced hot air convection MSc dissertation, pp. 75 (2009) (In Portuguese)
40. Seyed-Yagoobi, J., Wirtz, J.W.: An experimental study of gas-fired infrared drying of paper. Drying Technol. **19**(6), 1099–1112 (2001)
41. Wanyo, P., Siriamornpun, S., Meeso, N.: Improvement of quality and antioxidant properties of dried mulberry leaves with combined far-infrared radiation and air convection in Thai tea process. Food Bioprod. Process. **89**(1), 22–30 (2011)

Dynamic Drying Variables Evolution in Membrane Structure

Zawati Harun and Tze Ching Ong

Abstract A rigorous and comprehensive coupled mathematical model of mass, heat and dry gas transfer was implemented to analyze the variables variations in convective drying of porous materials. The governing system of nonlinear partial differential equations were derived where conservation laws were applied and implemented to finite element method in two dimensional system. Further, Skyline solver was used to capture highly nonlinear transient process. The characteristic of hygroscopic and nonhygroscopic materials on the drying variables variation was clearly distinguished in this works. The model was further improved in the application of multilayer membrane structure. Investigation of the dominant variables in this structure during the drying show that liquid diffusivity induced by capillary mechanism is dominant in the constant rate period (CRP) before vapour diffusivity caused by diffusion of vapour pressure and bulk flow gas take place in first falling rate period (FRP1). Subsequently, when drying reaches the second falling rate period (FRP2), the bound water mechanism is activated for hygroscopic zone whereas drying is almost accomplished for nonhygroscopic materials. At the shift to a multilayer structure system, top hygroscopic layer exhibits a slower drying rate due to higher water retention associated with its material characteristic compared to nonhygroscopic layers. The model is able to predict proper results with reasonable accuracy at any times. Knowledge gained from this study can be used to assist with the optimization of a given dryer design during drying process of ceramic membrane fabrication.

Keywords Diffusivity · Nonhygroscopic · Hygroscopic · Bound water · Convective drying · Membrane · Ceramic

Z. Harun · T.C. Ong (✉)
University Tun Hussien Onn, Parit Raja, 86400 Batu Pahat, Johor, Malaysia
e-mail: alex_ongtc@yahoo.com

J.M.P.Q. Delgado and A.G. Barbosa de Lima (eds.),
Drying and Energy Technologies, Advanced Structured Materials 63,
DOI 10.1007/978-3-319-19767-8_7

1 Introduction

Drying phenomena involve simultaneous heat, mass and gas transfer is a complex process since it combines lots of variables that evolve concurrently with time. The process normally consists of unsaturated flow of liquid within the porous solid, vapour flow within the porous solid, the liquid vapour phase change and convective-diffusion transfer of vapour from the surface to the surroundings [1]. Hence, typical drying stages are normally categorized as CRP, FRP1 and FRP2 [2–4] which aforementioned processes take place. It should be noted also that drying always accompany with shrinkages mechanism that generated from non-uniform stresses such as pressure gradient from the flow of capillary flow in CRP, macroscopic pressure gradient of escaping gasses during FRP and different thermal expansion of ceramic due to temperature gradient [5], that causes warping or cracking. The flow of liquid, the macroscopic pressure gradient, and the temperature gradient is controlled by the drying rate and rate of drying is typically controlled by external conditions [5]. Thus, understanding the drying process variables variations are essential towards controlling the drying rate.

Both hygroscopic and nonhygroscopic material has different drying behavior and mechanism. Drying for nonhygroscopic porous materials only free water or unbound water contributes to the flow of water inside the pores. When moisture content reaches an irreducible level, drying process is almost accomplished for nonhygroscopic material. The portion of water inside a very fine capillaries pore is defined as irreducible water content or bound water [6]. Conversely, the hygroscopic material is strongly related to the movement of bound water at the irreducible stage, which can be described by the water flow along the very fine capillaries. However, when drying water content is higher than the irreducible water content, water will still exist as free water in the pores. Previous studies on drying of hygroscopic and nonhygroscopic have present different equations and formulations as well as the concept that had been derived [2–4, 6, 7]. Bound water movement is expressed in terms of the diffusion of sorbed water driven by a gradient in the chemical potential of the sorbed water molecules [7]. Stanish et al. [8] had derived a uniquely explicit expression for the bound water flux in hygroscopic material based on temperature and vapor pressure gradients. Meanwhile, Haghi [2] and Zhang et al. [6] revealed that the bound water transport mechanism only effective when saturation irreducible is exceeded. Although substantial literatures are available for drying in hygroscopic materials and nonhygroscopic materials, there is limited study on the comparison of drying variables variation between both materials.

The effects of multilayer membrane structures vary greatly with different characteristic of drying compared to single layer structure. Figure 1 depicts the typical general structure of a ceramic membrane. The top layer generally has fine and dense pore structure is used as a filter for separation process [9]. Meanwhile, the next three layers which are intermediate and bottom layer consist of mesoporous and macroporous ceramic materials respectively is fabricated in the form of porous structure to support the membrane itself and also to ease the permeation process [9].

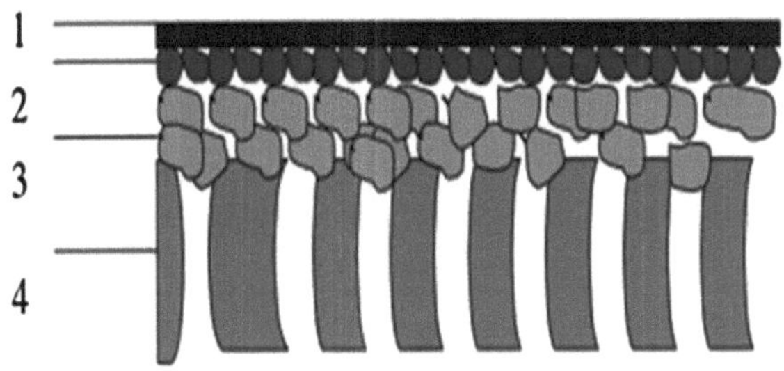

Fig. 1 Schematic representation of an asymmetric membrane (extracted from Li [9])

Therefore, the different drying characteristic is expected as attribute to different material behavior and properties. Hence, modeling the drying process of a multilayer structure system as a whole is crucial process to study the different variables evolution along the processes. Separate modeling for each material in drying process is much easier approach compared to construction of drying membrane structure comprising both materials of all periods of drying is necessary if we were to optimize the drying process using simulation results [10].

In this study, the implemented of governing equations of coupled mass, heat and gas transfer in two dimensional is an extension from previous related work [3, 11–13]. The transport mechanism of liquid by capillary action, vapor by diffusion and gas by vapor diffusion and bulk air flow in convective drying phenomena were taken into consideration in this model. In this present work, an improved model to accommodate hygroscopic material, which emphasized on bound water on later period (FRP2) of drying, was developed by referring to the previous work [2, 6]. Eventually, after instilled confidence in the proposed model in single layer structure, it was further extended to explore the drying evolution in the multilayer structure system.

2 Theoretical Formulation for Phase Transport

General mass and energy conservation law are used to define the mechanism of mass and heat transfers during the drying process in this model as a whole of multilayer ceramic membranes is described in this section. In the governing equations, the water is represented with the dissolved vapor and air and these are considered in the liquid phase, whilst the binary mixture of vapor and dry air is considered a gas phase. Capillarity effects are included for liquid phases, diffusion is also present for vapor and air. Gas transport and also heat transport are taken into account as these phenomena represent the drying process. Based on these physical principles, the governing equations are expressed in terms of three primary variables; pore water pressure (P_l), temperature (T) and pore air pressure (P_g).

$$\frac{\partial(\varphi\rho_l S_l)}{\partial t} + \frac{\partial(\varphi\rho_v S_g)}{\partial t} = -\nabla.(\rho_l V_l) - \nabla.(\rho_v V_v) - \nabla.(\rho_v V_g) - \nabla.(\rho_l V_b) \tag{1}$$

The water velocity, V_l and gas velocity, V_g can be easily derived from Darcy's law [2, 3] as:

$$V_l = -\frac{Kk_l}{\mu_l}[\nabla(P_l + Z)] \tag{2}$$

$$V_g = -\frac{Kk_g}{\mu_g}[\nabla(P_g)] \tag{3}$$

where Z is the vertical elevation from a datum. The vapor velocity by diffusion is defined as [14];

$$V_v = \frac{-D_{atm}v\alpha\theta_a}{\rho_v}.\nabla\rho_v \tag{4}$$

where D_{atm}, v, α, θ_a are the molecular diffusivity of water vapor through dry air, mass-flow factor, tortuosity factor and volumetric air term respectively. Rearranging the above equation according to the measured variables gives;

$$V_v = \frac{-D_{atm}v\alpha\theta_a}{\rho_v}\left\{\rho_o\frac{\partial h}{\partial P_l}\nabla P_l + (\rho_o + h\beta)\frac{\partial h}{\partial T}\nabla T + \rho_o\frac{\partial h}{\partial P_g}\nabla P_g\right\} \tag{5}$$

At shift to FRP2, the bound water flux is taken into account into the equation for hygroscopic materials. The bound water velocity is derived from Haghi [2] and Zhang et al. [6] as;

$$V_b = -D_b\nabla\theta_b \tag{6}$$

Substituting the above equation according to the measured variables gives;

$$V_b = -D_b\varphi\left(\frac{\partial S_l}{\partial P_l}\nabla P_l + \frac{\partial S_l}{\partial T}\nabla T + \frac{\partial S_l}{\partial P_g}\nabla P_g\right) \tag{7}$$

A mass balance to the flow of dry air within the pores of the material body dictates that the time derivative of the dry air content is equal to the spatial derivative of the dry air flux and is given as;

$$\frac{\partial(\varphi\rho_a S_g)}{\partial t} = -\nabla(\rho_a V_g - \rho_v V_v) \tag{8}$$

Heat energy transfer equations with the effect of conduction, latent heat and convection are considered and are given as below;

$$\frac{\partial((1-\varphi)c_p\rho_s + \sum_{i=a,l,v}\varphi s_i\rho_i c_i)}{\partial t} = \nabla(\lambda\nabla T) - \nabla(\rho_v V_v + \rho_v V_g)L - \nabla(\sum_{i=a,l,v}(T-T_r)\rho_i c_{pi} V_i) \quad (9)$$

In this analysis, the main assumptions involved in the formulation of drying model are

Solid, liquid and gaseous phases are considered, and these three phases are always in local thermodynamic equilibrium.
The fluid is present in liquid and vapour phases.
The movement of moisture in the porous skeleton is sufficiently slow so that in practice the temperatures of the liquid and vapour phases in the body are equal at coincident points.
The gas phase located in the voids is composed of an air and vapour mixture and the components follow the Ideal Gas Law.
Dimensional changes which occur within the material, due to a temperature or moisture content change, are comparatively small and will be ignored.
The matrix is rigid, homogeneous and isotropic in respective layer.
Darcy's law holds for the gas and liquid phases.
Gravity is important for liquid, but not for the gas phase.
Stresses developed are excluded.
Bound water movement only occurs in hygroscopic material and there is no interconnection capillary jump movement at the interface of hygroscopic and non hygroscopic material structure.

3 Thermodynamic Relationship

Coexisting liquid and vapour are assumed to be in local equilibrium within the porous. According to Kelvin's law, relative humidity is described by

$$h = \exp\left(\frac{P_w - P_g}{\rho_l R_v T}\right) \quad (10)$$

The vapour partial pressure can be defined as a function of local temperature and relative humidity as follows:

$$\rho_v = \rho_o h \quad (11)$$

where the saturation vapour pressure, ρ_o is estimated using the equation in Mayhew [15] with the saturated vapour density expressed as a function of temperature, which here has the form,

$$\frac{1}{\rho_o} = 194.4 \exp\left\{-0.06374(T-273) + 0.1634 * 10^{-3}(T-273)^2\right\} \tag{12}$$

The degree of saturation, S is an experimentally determined function of capillary pressure and temperature.

$$S_l = S_l(P_c, T) \tag{13}$$

Saturation is expressed in the equation below [16].

$$S_l = \frac{\theta - \theta_r}{\theta_s - \theta_r} = \left(\frac{1}{1 + (\alpha P_c(T))^n}\right)^m \tag{14}$$

where the parameters α, n and m are dependent on the porous material properties and these influences the shape of the water retention curve. Values for those parameters are defined in related work [17]. The permeability of water and gas are based on Muelem's model and are given by the formulation below as [18].

$$k_l(S_l) = \begin{cases} \sqrt{S_l}(1-(1-S_l^{\frac{1}{m}})^m)^2 & S_l > S_{irr} \\ 0 & S_l < S_{irr} \end{cases}$$

$$k_g(S_l) = \begin{cases} \sqrt{1-S_l}(1-S_l^{\frac{1}{m}})^m)^{2m} & S_l < S_{cri(g)} \\ 0 & S_l > S_{cri(g)} \end{cases} \tag{15}$$

4 Material Data

The sample used is referred as a very thin rectangular slab of length. Domains used for both study cases are illustrated in Fig. 2. Later, the slabs are assumed to be a porous medium that is homogeneous, isotropic and composed of solid phase, water and vapor phase, gas phase and dry air phase. The initial saturation was set to be S_O = 0.6 and the initial temperature T_O = 25 °C. The material properties and transport parameter used as data for simulation are listed in Table 1.

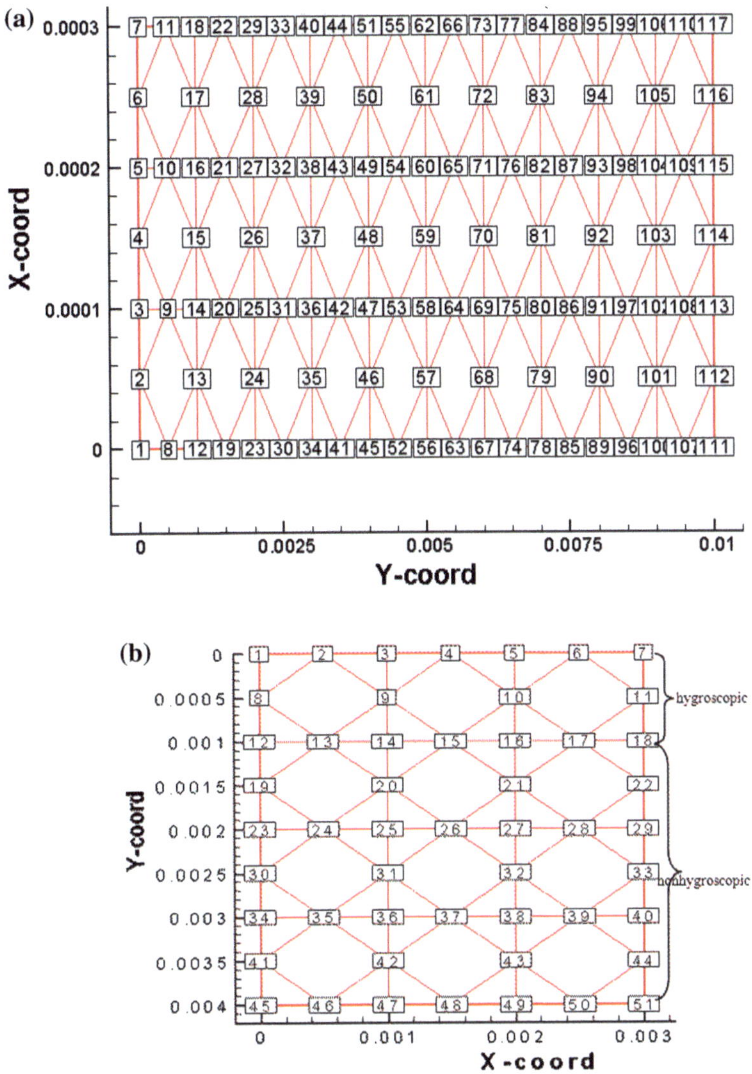

Fig. 2 Domain used: single layer (**a**); multilayer layer (**b**)

5 Boundary Condition

The general boundary condition is applied for convective mass and heat transfer.

$$J_m = h_m(P^v_\infty - P^v_s) \tag{16}$$

Table 1 Physical properties of ceramic body and transport parameter

Layer			Nonhygroscopic	Hygroscopic
Density of porous matrix	ρ_s	kg/m^3	2000	2600
Porosity	φ	1	0.2	0.17
Intrinsic permeability	K	m^2	10^{-17}	10^{-19}
Thermal conductivity of the porous matrix	λ	W/m K	1.8	1.8
Specific heat capacity of porous matrix	C_p	J/kg K	920	920
Critical saturation	S_{cri}	1	0.3	0.3
Irreducible saturation	S_{irr}	1	0.09	0.09
Equilibrium saturation	S_{eq}	1	–	0.05

$$J_T = h_c(T_f - T_\infty) \tag{17}$$

The boundary conditions of convective drying are applied to the top side of single layer at node 1–7 (refer Fig. 2a) and fluxes are applied to the side surface at node 1, 8, 12, 19, 23, 30, 34, 41 and 45 for multilayer structure (refer Fig. 2b) while the rest of the surfaces are impermeable and insulated for heat and moisture flow. Also the surface exposed fluxes assumed to be remain under atmospheric condition. The ambient temperature is set at 25 and 27 °C for single and multilayer cases respectively, with heat coefficient of 2 W/m^2 K for both cases and mass coefficient of 0.009 and 0.0035 ms^{-1} for single and multilayer structure respectively.

6 Solution of Governing Equations and Numerical Method

The coupled heat and mass transfer equations described above, in 2-dimensions can be written into a matrix form as follows;

$$[C(\Phi)\frac{\partial}{\partial t}\{\Phi\} = \nabla([K_{cx}(\nabla\Phi)]i_x + [K_{cy}(\nabla\Phi)]i_y)\{\Phi\} + R(\nabla Z) \tag{18}$$

where $\{\Phi\} = \{P_w, T, P_g\}$ is the column of unknowns; $[C]$, $[K_{cx}]$ and $[K_{cy}]$ are 3×3 matrices. Each element of the matrix is a coefficient for the unknown $\{\Phi\}$; i_x and i_y are the unit direction vectors. In order to discretize this simplified second order non-linear coupled partial differential equation, the finite element method is used. The Galerkin method was used to minimize the residual error before the application of Greens theorem, to the dispersive term that includes second order derivatives; this simplified combined equation set can be expressed in the following form.

$$\underline{K}(\Phi)\underline{\Phi} + \underline{C}(\Phi)\dot{\Phi} + \underline{J}(\Phi) = \{0\} \tag{19}$$

where,

$$\underline{K} = \begin{bmatrix} \underline{K}_{11}\,\underline{K}_{12}\,\underline{K}_{13}\,\underline{K}_{14} \\ \underline{K}_{21}\,\underline{K}_{22}\,\underline{K}_{23}\,\underline{K}_{24} \\ \underline{K}_{31}\,\underline{K}_{32}\,\underline{K}_{33}\,\underline{0} \end{bmatrix}$$

$$\underline{C} = \begin{bmatrix} \underline{C}_{11}\,\underline{C}_{12}\,\underline{C}_{13} \\ \underline{C}_{21}\,\underline{C}_{22}\,\underline{C}_{23} \\ \underline{C}_{31}\,\underline{C}_{32}\,\underline{C}_{33} \end{bmatrix}$$

$$\underline{J} = \begin{Bmatrix} \underline{J}_1 \\ \underline{J}_2 \\ \underline{J}_3 \end{Bmatrix}$$

$$\underline{\Phi} = \begin{Bmatrix} \underline{P}_{ws} \\ \underline{T}_s \\ \underline{P}_{gs} \end{Bmatrix}$$

$$\underline{\dot{\Phi}} = \begin{Bmatrix} \frac{\partial \underline{P}_{ws}}{\partial t} \\ \frac{\partial \underline{T}_s}{\partial t} \\ \frac{\partial \underline{P}_{gs}}{\partial t} \end{Bmatrix}$$

In which typical elements of the matrix are

$$\underline{K}_{ij} = \sum_{s=1}^{n} \int_{\Omega^e} K_{ij} \nabla N_r \nabla N_s d\Omega^e$$

$$\underline{C}_{ij} = \sum_{s=1}^{n} \int_{\Omega^e} C_{ij} N_r N_s d\Omega^e$$

$$\underline{J}_i = \int_{\Omega^e} K_{i4} \nabla N_r \nabla z d\Omega^e - \int_{\Gamma^e} N_r \underline{J}_i d\Gamma^e \quad (\mathrm{i},\ \mathrm{j} = 1, 2, \ldots., 4)$$

($\bar{n}$—outward normal vector to the boundary, Γ of the domain Ω).

The transient matrix and nonlinear second order differential equations above are then solved by using a fully implicit backward time stepping scheme along with a Picard iterative method which is taken into account for non-linearity.

7 Results and Discussion

The model proposed is verified with the computed results gained in Przesmycki and Sturmillo [19] modeling work. Figure 3 shows the variation of average saturation generated by proposed model and Przesmycki and Sturmillo [19] are in a close agreement. Also, it is obviously shows that the curve consists of three periods. In the next following paragraph, the drying variables such as pore water pressure, temperature and gas pressure evolution at different time will be discussed for each drying stage for both hygroscopic and nonhygroscopic material in a single layer case.

The first period presents almost a straight line from point A to B in Fig. 3 is called CRP. At this stage, free water will evaporate steadily and continuously. This is obviously shown by both plot of hygroscopic and nonhygroscopic materials of pore water pressure increased gradually when drying is at this stage (refer Fig. 4). The CRP lasted as long as the surface is supplied with liquid and depended strongly on the drying conditions and medium properties [2, 20]. The temperature increased slowly to the target temperature set (refer Fig. 5). As for the gas pressure in CRP, it

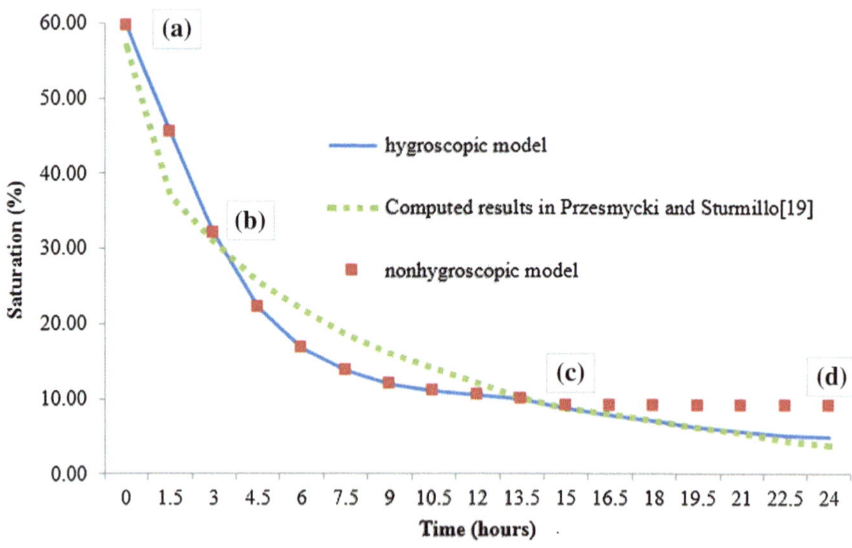

Fig. 3 Saturation distribution as a function of time

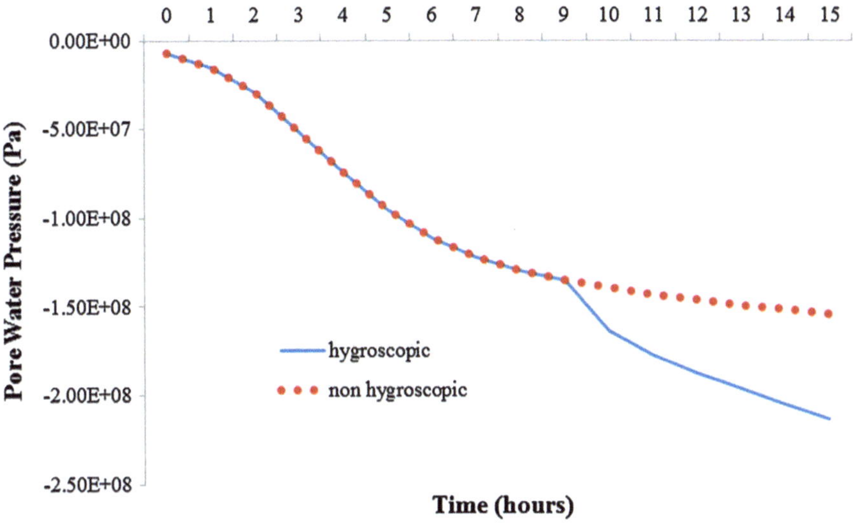

Fig. 4 Pore water pressure distribution as a function of time

stays constant or only show a slightly increased in value (refer Fig. 6) when vapour diffusion is less at this stage.

Eventually, when CRP is finished or when the water saturation reaches its critical free water value, which is 0.3 for most material [2, 3, 6]. The next stage is FRP1 and is shown by point B to C in Fig. 3. Pore water pressure shows a rapid

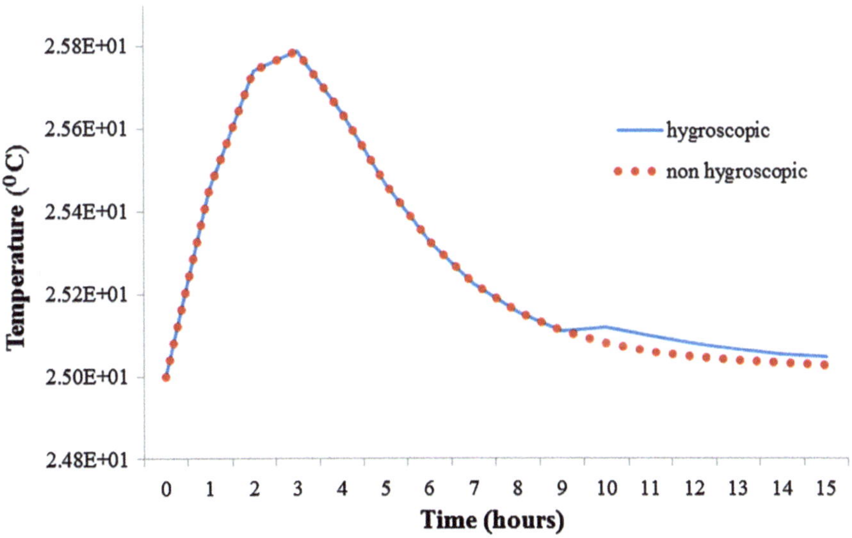

Fig. 5 Temperature distribution as a function of time

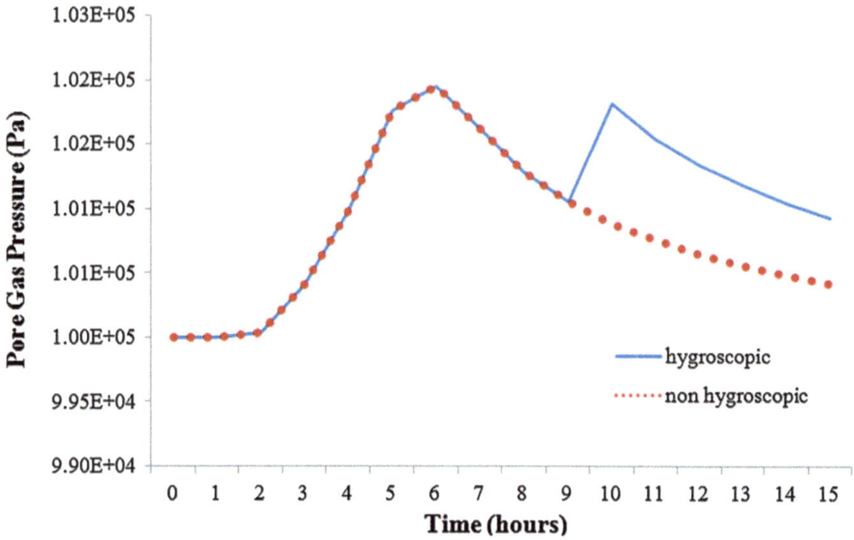

Fig. 6 Pore gas pressure distribution as a function of time

increased for both hygroscopic and nonhygroscopic as the water level is at the percolation threshold where the suction is greater at this stage (refer Fig. 4). Temperature profile also increases to its maximum value inside the body due to conduction before slowly decreases towards the ambient temperature (refer Fig. 5). Further decreases in saturation increased the volume fraction for gas phase. Hence the gas pressure will increase significantly at this stage as it is the dominant mechanism replacing the capillary mechanism earlier (see Fig. 6).

The last stage of drying is when water content exceeds 0.09 and shows the irreducible water value as shown in Fig. 3 from point C to D indicating FRP2. In nonhygroscopic materials, drying process at this stage is accomplished as pore water pressure, gas pressure and temperature started to decay towards the ambient condition. Conversely, for hygroscopic materials, this is a particularly important stage of the drying process where bound water will continue to be drawn from very fine and dried and brittle capillaries inside the porous structures. The bound water exist at this stage is strongly attached to solid matrix [2, 6] that shown by increasing of pore water pressure (refer Fig. 4). Driving force that represented by the gas phase or vapour pressure gradient [2, 6] that reflects higher gas pressure as shown in Fig. 6. It should be noted also temperature is slightly higher due to the conductivity of bound water inside a fine pore (refer Fig. 5).

In order to have a better understanding of the contributions of different transport mechanisms during drying, typical distributions of various components diffusivity for both nonhygroscopic and hygroscopic materials are presented in Fig. 7. In Fig. 7c, both types of materials show similar liquid diffusivities. When saturation is

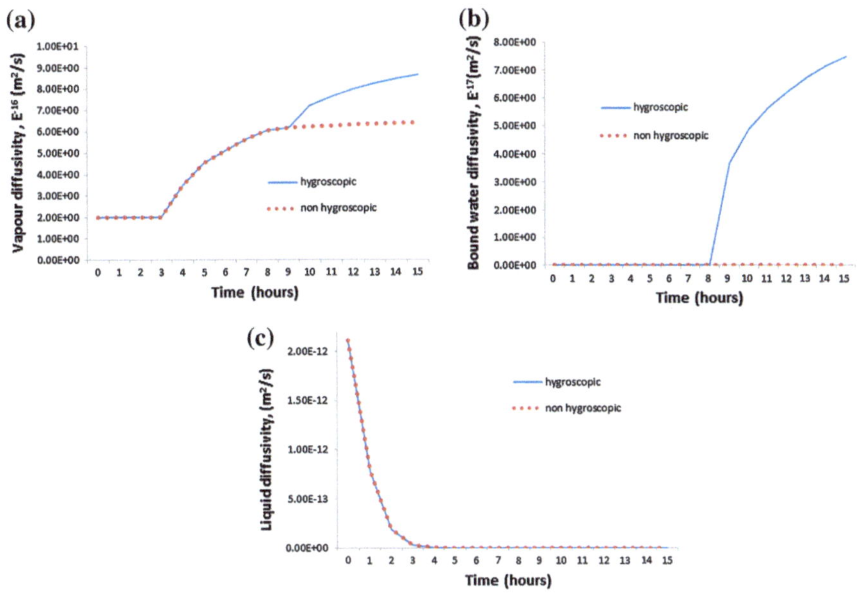

Fig. 7 Variations of diffusivity with time **a** vapour diffusivity, **b** bound water diffusivity, **c** liquid diffusivity

higher than saturation critical, liquid movement is dominant. This movement of water to evaporation front surface is maintained by capillary action. As saturation nears critical value, liquid diffusivity reduces sharply as evaporation front recedes from the surface as drying progress. Further drying when FRP1 shows zero liquid movement indicating capillary action is diminished. In the CRP, vapour diffusion (refer Fig. 7a) is in constant trend as the main mechanism of transport is a capillary action. When saturation reached the FRP1, vapour diffusion is increased significantly as saturation decreased and the volume fraction of gas phase increased. At irreducible saturation, the vapour diffusion is the only dominant mechanism of mass transfer for hygroscopic material. Bound water movement only exists in hygroscopic material as shown in Fig. 7b. The bound water transport mechanism will only effective when reached irreducible saturation. The movement of bound water in hygroscopic materials is also known as liquid moisture transfer near dryness or sorption diffusion [6] with the driving force only involves vapor transport mechanism [21]. The results gained in diffusivity study cases is in concordance with the results of Wang and Chen [22].

This model was further extended and modified to provide an insight into the drying process for the multilayer membrane structure system as a whole. Figure 8 demonstrates the saturation contour at the initial period of drying at 20 min and late stage of drying at 3.25 h. As been concluded from the previous results, the capillary mechanism is very profitable for moisture transport and flux of liquid being

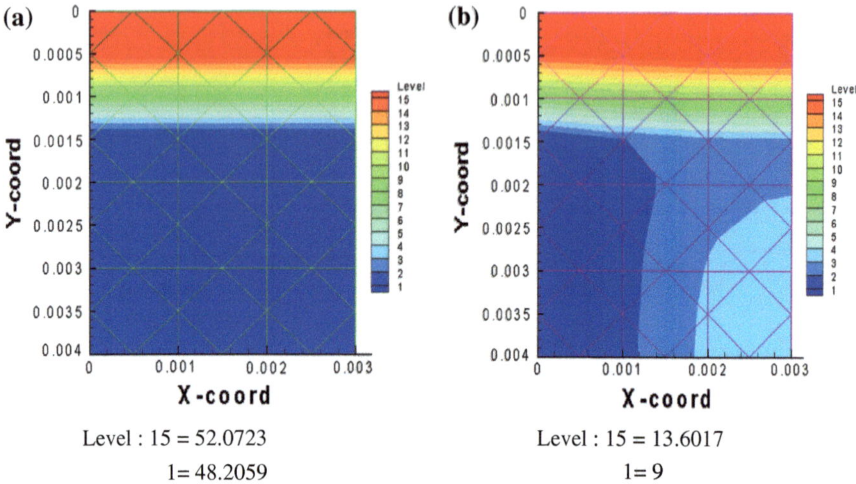

Fig. 8 Comparison of saturation contour at 20 min (**a**); 3.25 h (**b**)

generated to the boundary surface of the material. Hence, this effect is greatly shown in the late stage of drying as surface of the material is drier when compared to the interior parts. Also noted, hygroscopic layer exhibits higher water retention causes higher saturation at all times due to its material properties and behaviour.

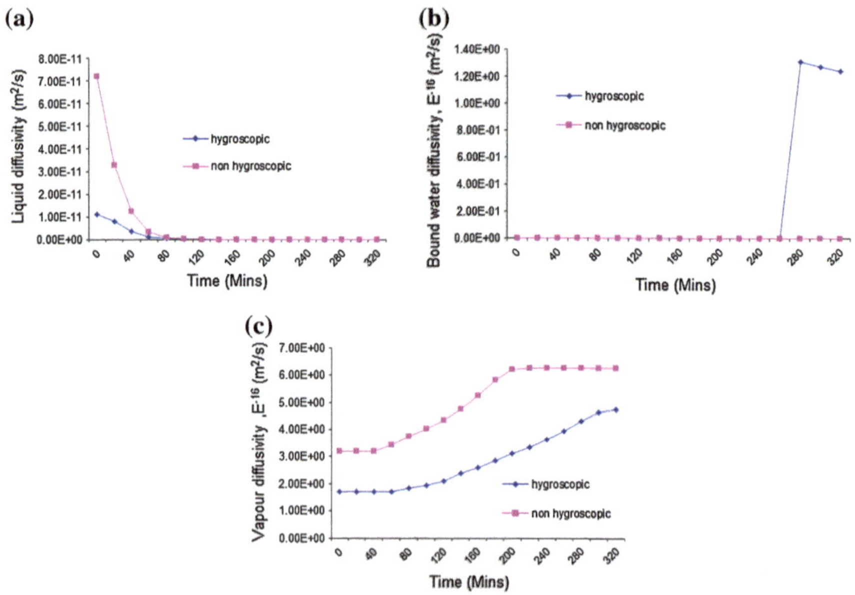

Fig. 9 Variations of diffusivity with time for multilayer system **a** liquid diffusivity; **b** bound water diffusivity; **c** vapour diffusivity

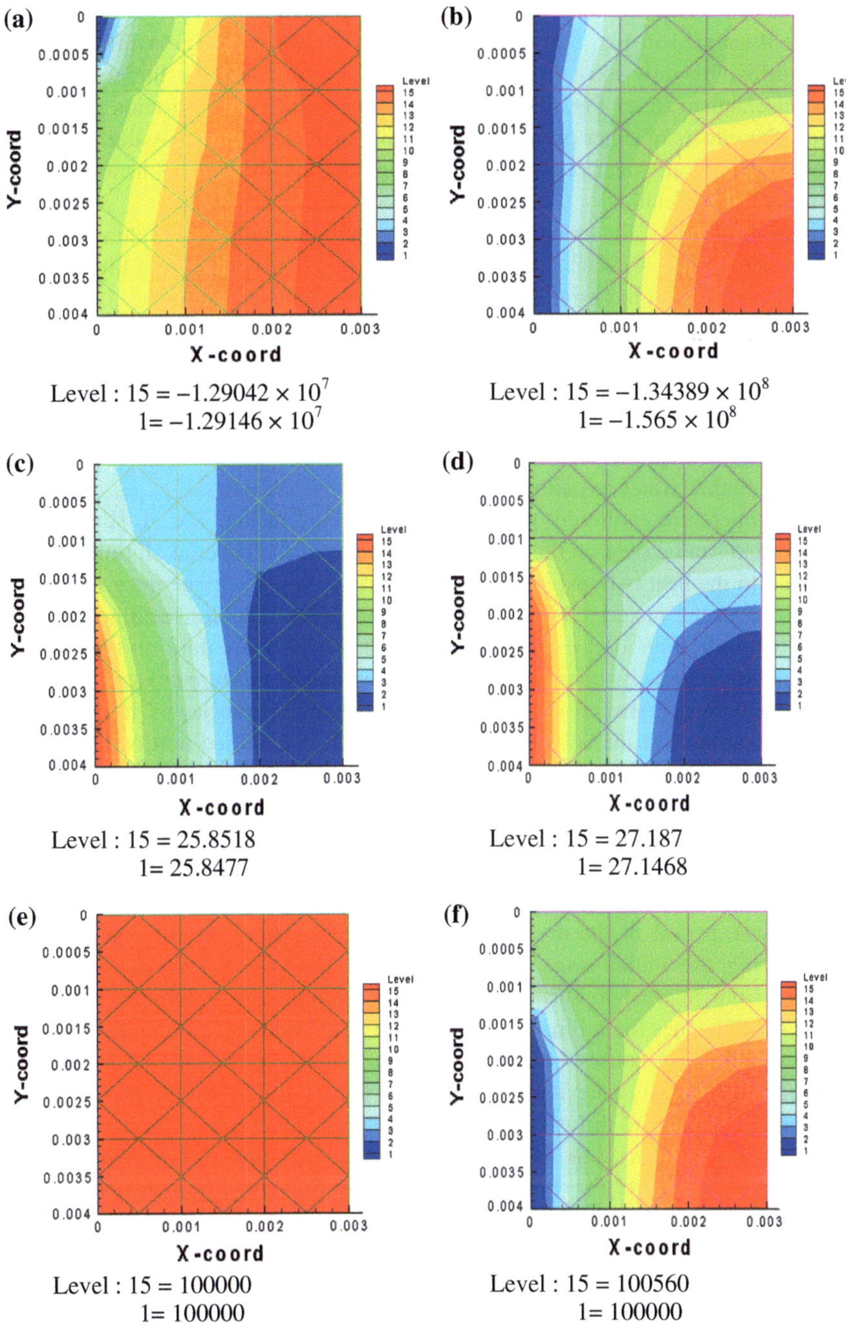

Fig. 10 Comparison of contour of: pore pressure water at 20 min (**a**) and 3.25 h (**b**); temperature at 20 min (**c**) and 3.25 h (**d**); gas pressure at 20 min (**e**) and 3.25 h (**f**)

This leads to higher liquid diffusivity for nonhygroscopic layer when drying is at CRP as illustrated in Fig. 9a. Eventually, when drying period is at FRP2, bound water diffusivity only exist in hygroscopic layer postulated the initial of bound water mechanism (refer Fig. 9b). Since the proper pattern of saturation contour is gained, the effects of other drying variables that evolve concurrently during drying period such as pore water pressure, temperature and gas pressure is next discussed and analyzed in detail.

Figure 10 shows the contour for pore water pressure, temperature and gas pressure at the start and late stage of the drying process. As has been illustrated in Fig. 10a, b, the most intensive part of the free mass transfer takes place on the side drying surface of the rectangular sample which is the boundary between the material and the external surroundings. When drying progress, the drying evaporation front recedes into the interior of the material causing higher capillary suction when moisture content is getting lesser. Figure 10a depicts hygroscopic layer has higher pore water pressure at all times when compare with nonhygoscopic layers causes by higher water retention within the pores.

The temperature evolution during drying is shown in Fig. 10c, d. Higher temperature also can be seen on the side drying surface of the material all times, which attributed to the heat is transferred from the ambient to the surface material then to the interior of the material relies mainly on conduction and convection [10]. It is evident from the Fig. 10c, d that hygroscopic layer has more uniform temperature profile mainly causes by constant conductivity at higher moisture content inside the pores at all times. Also, it can be noticed, nonhygroscopic layers have slightly higher temperature at the side drying surface which is the dried portion. As drying process, the dry zone extends gradually and temperature increases within the material. Eventually, the temperature in the material is in equilibrium with the surrounding temperature when drying is accomplished [10].

Figure 10e, f shows the gas pressure contour during the early and late stage of drying process. According to Fig. 10e, the gas pressure slowly builds up at the constant rate period of the drying process where moisture content is high inside pores during the initial stage and capillary mechanism is the main supply of moisture from internal to surface material for evaporation to happen. When the drying stage reaches the falling rate period, pressure builds up more drastically due to vapour diffusion causes by phase transition change of water to vapour [20]. Subsequently, gas pressure receded to atmosphere condition as drying is almost accomplished. Vapour diffusivity results reveal that nonhygroscpic layers show higher vapour diffusivity as faster drying rates are dominant compared with the hygroscopic layer (refer Fig. 9c). However, vapour diffusivity in hygroscopic layer still propagates when it is already ceases for nonhygroscopic layer due to bound water mechanism which vapour diffusion is the driving forces.

Overall, the contour data represent a similar trend with the study by Kowalski [23, 24]. Also noted, the model proposed which correlation between the properties of the material and drying period. Thus this proposed model is an effective tool to obtain a better understanding of the underlying physics for multilayer material of the drying process at a fundamental level. The knowledge gained from this study can be used to investigate the defects that might occur associated with different shrinkages of materials for future study.

8 Conclusions

The evolutions of various variables associated with the drying process for hygroscopic and nonhygroscopic were investigated for both single layer and multilayer structure. For single layer case, both materials show similar configuration during CRP and FRP1 as capillary is dominant in CRP, which is onset to vapour diffusion during FRP1. The significant effects occur when drying reaches FRP2 where bound water mechanism exist in hygroscopic materials whereas drying is almost accomplished for nonhygroscopic material. Eventually, it also results in higher pore water pressure, temperature and gas pore pressure of the associated variables. Thus, this study clearly shows the different mechanism evolves in the drying process for both materials. Subsequently, a further case study involves multilayer sample postulate hygroscopic layer has higher moisture content, pore water pressure and more uniform temperature attributed to its material properties. Eventually, gradient of pore gas pressure is higher in nonhygroscopic layers causes by faster drying rates. Also noted, side drying surface exhibits maximum temperature and pore water pressure as extremely moisture extraction occurs at this position. Thus, this also causes minimum saturation at this position. Lastly, this model is considered to be a useful tool for simulating the process of single layer and multilayer convective drying process due to its accurate prediction of variables during all periods of drying process at all times and conditions.

Acknowledgments Authors thank to the financial support by University Tun Hussien Onn Malaysia (UTHM) and Ministry of Higher Education Malaysia (MOHE).

References

1. Hall, C., Hoff, W., Nixon, M.: Water movement in porous building materials—VI. Evaporation and drying in brick and block materials. Build. Environ. **19**(1), 13–20 (1984)
2. Haghi, A.K.: Transport phenomena in porous media: a review. Theor. Found. Chem. Eng. **40** (1), 14–26 (2006)
3. Harun, Z., Gethin, D.T.: Drying(consolidation) porous ceramic by considering the microscopic pore temperature gradient. Appl. Mech. Mater. **147**, 210–214 (2012)
4. Harun, Z., Nawi, N.M., Batcha, M.F., et al.: Modeling of layering ceramic shell mould. Appl. Mech. Mater. **232**, 548–552 (2012)
5. Ring, T.A.: Fundamentals of Ceramic Powder Processing and Synthesis. Academic Press, San Diego (1996)
6. Zhang, Z., Yang, S., Liu, D.: Mechanism and mathematical model of heat and mass transfer during convective drying of porous materials. Heat Transf. Asian Res. **28**(5), 337–351 (1999)

7. Siau, J.F.: Chemical potential as a driving force for nonisothermal moisture movement in wood. Wood Sci. Technol. **17**, 101–105 (1983)
8. Stanish, M.A., Schajer, G.S., Kayihan, F.: A mathematical model of drying for hygroscopic porous media. AIChE J. **32**(8), 1301–1311 (1986)
9. Li, K.: Ceramic Membranes for Separation and Reaction. Wiley, Chichester (2007)
10. Kowalski, S.J., Rybicki, A.: The vapour–liquid interface and stresses in dried bodies. Transp. Porous Med. **66**, 43–58 (2007)
11. Harun, Z., Ong, T.C., Ahmad, R.: Drying comparison of nonhygroscopic and hygroscopic materials. Appl. Mech. Mater. **465**, 637–641 (2014)
12. Harun, Z., Ong, T.C.: Material parameters sensitivity in modeling drying of porous materials. Adv. Mater. Inf. Technol. Process. **87**, 91–98 (2014)
13. Harun, Z., Ong, T.C., Fauzi, A.F.: Simulation of drying for multilayer membranes. J. Teknol. **69**(9), 107–115 (2014)
14. Kanno, T., Kato, K., Yamagata, J.: Moisture movement under a temperature gradient in highly compacted bentonite. Eng. Geol. **41**(1–4), 287–300 (1996)
15. Mayhew, Y.R., Rogers, G.G.C.: Thermodynamic and Transport Properties of Fluids. Blackwell, Oxford (1976)
16. Genuchten, M.V.: A closed-form equation for predicting the hydraulic conductivity of unsaturated soils. Soil Sci. Soc. Am. J. **8**, 892–898 (1980)
17. Harun, Z., Ong, T.C., Ahmad, R.: Estimation of water desorption in drying of membrane structure system. Adv. Mater. Res. **1004–1005**, 405–408 (2014)
18. Baroghel-Bouny, V., Mainguy, M., Lassabatere, T., et al.: Characterization and identification of equilibrium and transfer moisture properties for ordinary and high-performance cementitious materials. Cem. Concr. Res. **29**, 1225–1238 (1999)
19. Przesmycki, Z., Strumillo, C.: The mathematical modelling of drying process based on moisture transfer mechanism. Drying **85**, 126–134 (1985)
20. Perré, P., Remond, R., Turner, I.W.: Comprehensive drying models based on volume averaging: background, application and perspective. In: Tsotsas, E., Mujumdar, A.S. (eds.) Modern Drying Technology, vol. 1. Wiley VCH Verlag GmbH & Co. KgaA, Weinheim (2007)
21. Defraeye, T., Blocken, B., Derome, D.: Convective heat and mass transfer modelling at air–porous material interfaces: overview of existing methods and relevance. Chem. Eng. Sci. **74**, 49–58 (2012)
22. Wang, Z.H., Chen, G.: Heat and mass transfer during low intensity convection drying. Chem. Eng. Sci. **54**(17), 3899–3908 (1999)
23. Kowalski, S.J.: Control of mechanical processes in drying. Theory and experiment. Chem. Eng. Sci. **65**(2), 890–899 (2010)
24. Kowalski, S.J.: Continuous thermohydromechanical model using the theory of mixtures. In: Tsotsas, E., Mujumdar, A.S. (eds.) Modern Drying Technology, vol. 1. Wiley VCH Verlag GmbH & Co. KgaA, Weinheim (2007)

Technical Innovation in Dehydration Process for Wine Quality

B.J. Teruel, R.C.R. Tinini, F. Mencarelli, R.A. Oliveira and W.E. Santiago

Abstract The grapes dehydration to make wine, in controlled environment and with variance in temperature and humidity, began to be studied with greater emphasis recently for some grape varieties, due to the proven effects on the higher concentration of phenolic compounds and the positive change in aroma. This chapter focuses on process control in with particular emphasis in instrumentation, applied technologies and main quality results.

Keywords Temperature · Weight loss · Phenolic compounds · Soluble solids · Control process

1 Introduction

Among the products obtained from the grape juice processing, wine is a widely consumed beverage in the world. The production is supported by a past history of over 6000 years, and many studies have shown that a rational use of wine proved health benefits and cardiovascular disease prevention, reducing the likelihood of diabetes and liver disease. The world wines production is concentrated mainly in European continent (Italy—17.5 %; France—16.2 %; Spain—13.6 %), representing 47.31 % of world production, followed by the United States, with production 2,653,700.00 liters, representing 10.5 % of world production. This domain production in Europe is caused by the fact that wine is considered a part to the meal. It is justified by the carbohydrate,

B.J. Teruel (✉) · R.C.R. Tinini · R.A. Oliveira · W.E. Santiago
College of Agricultural Engineering, State University of Campinas (FEAGRI/UNICAMP),
Av. Cândido Rondon, 501, Barão Geraldo, 13083-875 Campinas, São Paulo, Brazil
e-mail: barbara.teruel@feagri.unicamp.br

F. Mencarelli
Department for the Innovation Biological Agrifood and Forest Systems,
University of Viterbo, Via de Lellis 01100, Viterbo, Italy
e-mail: mencarel@unitus.it

J.M.P.Q. Delgado and A.G. Barbosa de Lima (eds.),
Drying and Energy Technologies, Advanced Structured Materials 63,
DOI 10.1007/978-3-319-19767-8_8

vitamins and minerals presence in their constitution, with a water content of 80–85 %. For example, per capita comsumption shows this domain: in France, it is 46 l, Italy, 42.1 l, Portugal, 41.8 l and Switzerland 38.2 [1–3].

The Brazilian situation is somewhat different: the vineyard area takes little emphasis on the world stage, representing only 1.3 % of the acreage in the world. However, shows significant growth between 2007 and 2010 (almost 7 %), which has shown the Brazilian production potential. However, Brazil is also one big wines importer and processed products, which for wine sector amounted to 155.7 million in 2011, with an increase of 4.9 % compared to 2010. The Brazilian wine industry is still concentrated in small farms generally with family productions. This fact indicates that the wine industry is gaining importance for the country's development, creating jobs, starting new investments in larger enterprises for a table grapes production and processing, production increased between 2010 and 2011 by 13 %. The main grape producer states is Rio Grande do Sul, followed by Santa Catarina, Paraná, São Paulo and Minas Gerais and other considered emerging states are Pernambuco and Bahia [3–5].

With the advance of globalization and internationalization of markets, the Brazilian wine agribusiness has undergone a major expansion, as well as the requirements of safety standards and international quality (seals and origin certificates quality and safety, according to product features or region where it is produced).

The grape juice industry sector is not indifferent to the growth suffered by the grape production in the country. Data from ASBRASUCO [6], show that the grape juice participation is the highest one for Brazilian consumers, increasing from 12 to 28 % between 1999 and 2011. This advance was driven basically by the orange decreasing, which had reduced their participation of 48–13 % in the same period. About 50 % of the common grape production in Brazil are designed exactly for the production of juices, demonstrating the importance of the sector in the sustainability of the primary sector, which is engaged in viticulture [6].

Brazil consumes the equivalent of 3.2 liters per capita of fresh juices, or 618 million liters, still far from soft drinks market, which is 17 billion liters per year, or compared to Mexico, where consumption is 17.3 l. However, stands out the growth of this quota for grape juice, which now owns 28 % of consumer preference for all natural juices. Emphasizes yet the of nectar importance with sweet flavor and easy acceptance among consumers. This increase is due to two reasons: the increasing purchasing power in Brazil and concerns of people with a natural diet and healthy [6]. The current scenario indicating the increased consumption of wine and grape juice in the world, and particularly in Brazil, is conducive to the development of research related to quality of Viticulture derivatives sector, in the framework of the Precision Agriculture and Agribusiness.

2 State of Art

2.1 Theme Outlook

In developing countries, the post-harvest losses of fruits exceed 20 %. This fact is intrinsically associated to lack of appropriate technology during handling, transportation and storage, which are required for preservation and transformation into good quality products.

In recent decades, researches in fruit dehydration area have been widely expanded in search of products with minimum change in their sensory and nutritional characteristics. The fruit drying operation is one of the simplest and most cost-effective methods for food preservation.

The advantages of drying process are numerous. Among them, we can cite: the simplicity for preservation of the product; stability of aromatic components at room temperature for extended periods; protection against enzymatic and oxidative degradation; reduction of mass; energy savings by not requiring refrigeration and product availability during off-season periods.

By definition, drying is the removal of water from the biological material by evaporation. During drying process, it is indispensable the supply of heat to evaporate moisture from material and, also, a moisture drain (a sink) to remove water vapor formed in the material surface.

When placed in the dryer, the biological material is exposed to a different environment temperature (hotter than material), creating a heat transfer from the hot source for the wet material. Besides, the evaporation of water occurs, concomitantly. The water vapor partial pressure deficit between hot environment (hot air) and the product surface causes a mass transfer from the product to air, and so the vapor will be drawn from material.

If the water is not bound (physical and/or chemical connection) in solid structures, it is characterized as free water, and the energy involved in the process is related to the latent heat of vaporization. And, if the water is bound, the energy required for evaporation is greater.

During drying, the evaporation of water occurs on the surface of material, which has been transported from the inside of solid material. The most important transport mechanisms are liquid diffusion, vapor diffusion and flow of liquid and vapor.

Drying process of biological materials is based on the property of desorbing (lose) or adsorbing (gain) moisture from surrounding air, called hygroscopicity. For a given temperature, the moisture content of biological materials tends to come into equilibrium with the relative humidity. This phenomenon occurs due to the relationship between water vapor pressure of organic material and the water vapor pressure of environmental air. If the vapor pressure of organic material is lower than that present in air, the material tends to adsorb humidity (adsorption); in the reverse situation, the material will desorb moisture to the air (desorption). This last case consists on the principle used to dry biological material (drying process).

The initial and final (equilibrium) moisture content in the material, the relationship between water and solid structure and mass transport of water from inside the material to its surface support drying phenomenon.

Therefore, drying phenomenon cannot be generalized to biological material because they have characteristics and properties that can change intensely during drying.

The efficiency of drying process is related to the quality of final product, i.e., the desirable characteristics of resulting product.

A product that is exposed to drying process has to be qualified for a subsequent process or to be available to consumer market. In the case of grapes drying, it is intended to reduce its moisture content and to concentrate other compounds, modifying organoleptic characteristics and improving fermentation process.

2.2 *Partial Dehydration of Grapes to Produce Juices and Wines*

The drying, withering and dehydration are different terms which are used to explain the water removal from grape the declination of these terms depending on process [23]. Dehydration of grapes for wine making, on the inside and with the variation of temperature and humidity, began to be studied with greater emphasis recently for some varieties tested due to the effects concentrations of phenolic compounds increasing and reducing the amount of water in the must (Fig. 1).

Dehydration postharvest changes consistent results in post-harvest, thus obeying a positive development of the flavor of the wine and grape. Besides facilitating the transport, storage and microbiological stability, the dehydration process of agricultural products causes physical, chemical and organoleptic changes; therefore, it must be performed in a controlled manner and meet the limits established to not affect the quality [7]. The reduction of the moisture content in fruits for processing, whether for juice, pulp and/or concentrates, causes an increase in the concentration of soluble solids [8, 9]. When the grapes are used for winemaking, it is desirable the highest possible concentration of soluble solids and phenolic compounds. Phenolic compounds, as well as anthocyanin's and flavonoids, provide the sensory characteristics of the wine, and they are indicated as beneficial in the prevention of cardiovascular diseases [10–12].

According to the study of [13], the concentration of soluble solids in the must is the key factor so that, during winemaking, a significant amount of alcohol is produced. When the concentration of soluble solids is low, it becomes difficult, or even impossible, to obtain table wines with alcohol levels according to the required by the Brazilian law, which must be between 8.6 at 14 %, according the National Institute of Metrology, Standardization and Quality Industrial-Inmetro.

Studies developed by [14] show that the shriveling of grapes in the vineyards, caused by loss of water, contributed to significant changes in the concentration of

Fig. 1 **a** Automatized chamber to dehydrated grapes; (**b**, **c**) traditional methods applied in Italy to dehydrate; **d** grape appearance during the process [30]

total soluble solids (TSS), in addition to the increase in the concentration of phenolic compounds in the skin. Later studies on the application of partial dehydration at temperature below ambient began by using a cold chamber intended for this function, the dehydration process may take up to 45 days [15]. While in some countries with wine tradition the partial dehydration of grapes is an important methodology for achieving quality table wines [16], in Brazil it is common in many wineries the addition of sugar to the must, a process called chaptalization, in order to improve the alcohol content; however, this process results in low quality wines, lower market value and higher production costs. Results obtained from several studies have proven the positive effect on wine produced from grapes that have undergone partial dehydration, supported by the increased concentration of sugar, phenolic compounds and aromatic compounds [17–19, 28].

The biochemical changes that occur in the grapes during the wilting are induced by the endogenous metabolism of grapes and results in modification of the quality of wine. However, drying at ambient conditions is time consuming and brought other problems such as contamination by fungi and bacteria. The study of the drying process at low temperatures (10, 15, 20 and 25 °C) and relative humidity range from 53; 75 and 95 % significantly reduced the time required for water evaporation, achieving higher concentrations of sugars, decreasing the risk of *Botrytis cinerea*, which affects the wines quality.

The results obtained by [15] and [28] ensure that this procedure can enable cost reduction with the addition of sugar in the must. However, partial dehydration of grapes is still a procedure rarely used in Brazil, being commonly seen in traditionally wine producing countries [16]. However, the methodology practiced in many of these countries does not have any technique or tool that allows the control of environmental conditions and ensures the asepsis of the product.

The International Code of Oenologist Practices [20] states that grapes for winemaking, when subjected to drying processes, lose the maximum of 20 % of their initial moisture content and that the potential alcoholic content does not exceed 2 % of the volume. Dehydration above 20 % can cause physiological damage to the product, compromising its use in the must production.

Surveys with fruit for industrial manufacturing and processing of juice, pulp and concentrates show that after reduction of the moisture content, these can achieve greater amount of soluble solids, higher concentration of chemical components and increased microbiological stability [8].

Higher ambient temperature with low relative humidity in a drying chamber (40 °C and relative humidity of 20 %) the process is stopped when the sugar content is about 260 g L^{-1} and was observed that the concentration of the compound varieties had flavor and aromatic improvements in the must most [31].

In Brazil, grapes dehydrated under controlled conditions, using low temperature and relative humidity [32, 33] (T = 7 °C and RH = 35 %) and airflow of 12 m^3 s^{-1}. It was observed a slow process for 42 days to weight loss approximately 1.5 kg H_2O kg of dry $matter^{-1}$.

This results showed than dehydrated grapes suffer significant changes in the concentration of sugars, but with the control process such changes, is possible contribute to quality of the final product, without modified of the volatile phenolic compounds and enzyme activity [21, 22]. However, there is the need to control the variables affecting the process of heat and mass transfer (temperature, airflow, humidity) so that benefits are achieved both from a physiological point of view as food safety by eliminating the risk of fungal growth and thus pollution, for example, ocratoxin [15, 23].

With the advancement of science and technology, are proposing new methods of dehydration, the equipment development and drying systems (With the advancement of science and technology, have being proposed new methods of dehydration, equipments and drying systems.). Methods other than spray drying, sublimation, vacuum or forced air, there is still osmotic dehydrationmethod has been widely used in the transformation and processing of certain tropical fruits (About the drying methods, still has spray drying, sublimation, vacuum or forced air, besides the osmotic dehydration method what has been widely used in the transformation and processing of certain tropical fruits) [9, 24]. The difficulty in determining the optimal harvest point, with adequate concentration of soluble solids in the must, it is a challenge.

In Brazil, industrially processed grapes for wine production, generally, not submit to any form of heat treatment aiming for improvements in the quality of the wine produced. The difficulty in determining the optimal harvest point.

3 Instrumentation and Process Control in Grapes Partial Dehydration

The previous research results shows the feasibility of partial dehydration in grapes, to increase the soluble solids content and polyphenols concentration, without compromising the product quality, adding value, wine or juice and opening a perspective contribution to the wine industry. But, it is still necessary to further develop research to improve the technology and the possible transfer to the sector.

The automatic control and measurement of the involved variables in the process (temperatures, relative humidity and airflow), allows the supervision in real time the conditions of dehydration, looking for improvements in the final product heat and mass transfer.

The partial dehydration of grapes after harvest aimed at winemaking has been shown to be a process that brings increased concentration of sugar and phenolic compounds in the must, which affects the quality of the wines produced. However, the works developed so far have studied the process for temperatures up to a maximum of 25 °C and air velocity lower than 1 m s^{-1}.

The system described by [25–27] is instrumented with sensors for air temperature and relative humidity and mass measurement (Fig. 2b). The system instrumentation was already carried out in previous works of [29]. The temperature sensors are of the Pt 100 type (Range = 0 at 100 °C; model TR106; 4–20 mA; Accuracy = ±0.2 %); the ones to measure the relative humidity are of the RHT-WM type with compact electronic module and transmitter of values (FM = 0 at 100 % UR; 4–20 mA; Accuracy = ±1.5 %); and to measure mass, was used a weighing system comprising a load cell, model PW12C3–IMB (50 N (50 kgf), sensibility of 2 ± 0.1 % mV V^{-1}).

The application for the real-time monitoring was developed in graphical environment of Labview® programming software (National Instruments). The information related to sensing instruments is integrated into a central processing and data acquisition unit according to the diagram in Fig. 3. Data acquisition of temperature and relative humidity was carried out by the data acquisition board (PCI-NIDAQ 6229) coupled to a connector block (CB-68LP) both from National Instruments. This board allows to input analogical values as temperature and relative humidity expressed between 4–20 mA and to output a voltage that acts on the frequency inverters connected to compressor and exhaust fan which an vary between 0 and 10 V depending on the desired cooling efficiency and air velocity. Digital data of electrical meters and weighing system are transmitted to the microcomputer systems via Modbus protocol through the RS485 serial port, thus enabling data to

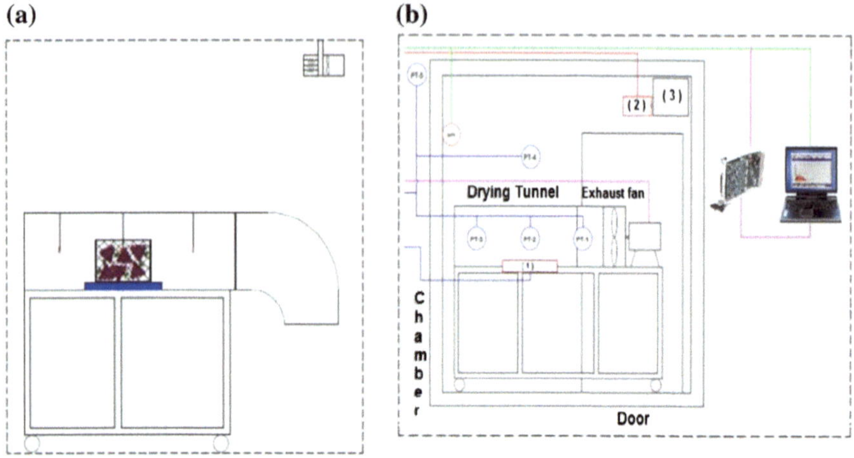

Fig. 2 **a** Schematic of the structure of forced-air drying; **b** schematic of the system instrumentation. (*1*) Scales; Resistance (*2*); Evaporator (*3*)

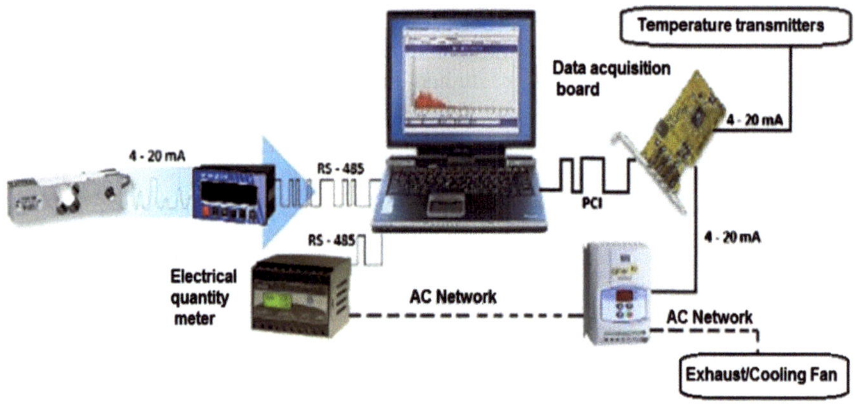

Fig. 3 Instrumentation diagram

be read and stored. All signals obtained with instrumentation system, after being processed by the computer, are displayed in real time on the application of supervision and are available as a source of information. Then, it supports to the decision-making related to changes in the parameters responsible for process kinetics, such as temperature and velocity of drying air.

The application for real-time monitoring resulted in development of an application called SECAGEM DE UVAS (DRYING OF GRAPES), which has the purpose of collecting, processing and storing data from the sensors installed in drying chamber, in addition to enabling the operator to make decisions on possible changes in the process.

The application DRYING OF GRAPES is available to the user in graphical and numerical form by the computer screen, showing data in real time. It also allows the user to control and determine the performances of some experimental physical system equipment. The human machine interface (HMI) also includes the monitoring of internal conditions of drying chamber, in which parameters such as temperature and relative humidity are exposed and monitored in real time by the operator by numerical and graphical indicators (Fig. 4). In this window is possible see the numerical indicator sampled by the weighing system (mass), the graphical of the mass loss during the course of the drying time, the variables control at the end of the process and set point.

The HMI overall supervisory system is shown in Fig. 4. As described above, this system is programmed to read the signals of the temperature sensors, relative humidity, mass and power consumption, in addition to acting on the operation of the compressor motors and exhaust fans, the operating frequency of the exhaust and the level of heat load dissipated by the resistances. The application collects the signals from the load cell at a rate of approximately 120 samples per second, forming an average result every minute; for the other sensors, the rate is 100 samples per second, also forming an average result for every minute.

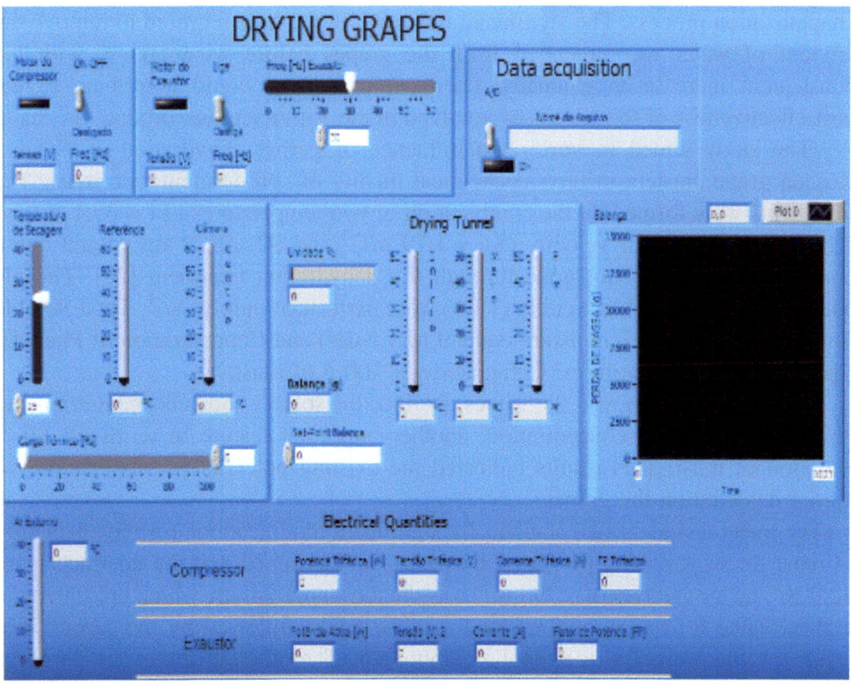

Fig. 4 HMI supervisory system

In order to analyze physical and chemical changes (concentration of total soluble solids (TSS) and phenolic compounds (CPC)) after partial dehydration of Niagara Rosada grapes submitted to two treatments combining two temperatures and one air velocity (T1 = 22.9 °C/1.79 ms^{-1} and T2 = 37.1 °C/1.79 ms^{-1}) and relative humidity of approximately 40 % [25]. The results show that it is possible, by modulating the psychometric and thermal parameters of dehydration process, to adjust the loss of water of grapes, with consequent increase in concentration of soluble solids and polyphenols. It also indicates that partial dehydration of grapes for winemaking can bring satisfactory results in the parameters of soluble solids and polyphenols, for the below and above room temperature.

The loss of water in the grapes was approximately 14 % and the drying process lasted between 20 to 50 h for the referred treatments T1 and T2, respectively. It was experimentally and statistically verified that the treatments promoted significant increase in TSS and CPC; however, for CPC at the temperature of 37.1 °C, the increase accounted for approximately 29 %, whereas, for the temperature of 22.9 °C, it was only 5 %. For TSS, the increase was on average 14.4 ± 3.9 % between both treatments.

The authors could conclude that the combination of forced air and temperature of 37.1 °C, it is possible to accelerate the process drying of grapes, providing increased concentration of phenolic compounds and total soluble solids, thus reaching the alcohol content in must required for the production of wines without chaptalization process. The significant reduction in the time required to remove the amount of water recommended from the grapes before winemaking, with the consequent increase in evaluated chemical properties, could contribute to the cost-effectiveness of the processes involved in the Vitiviniculture production chain.

The study aimed to verify the influence of partial dehydration of Niagara Rosada grape clusters in physicochemical quality of pre-fermentation must. The treatments were formed by the combination of two temperatures (T1 = 37.1 °C and T2 = 22.9 °C) two air velocity (S1 = 1.79 ms^{-1} and S2 = 3.21 ms^{-1}) and one a control (T0) that has not gone through the dehydration treatment [26]. The pH analysis, Total Titratable Acidity (TTA) were performed in mEq L^{-1}, Total Soluble Solids (TSS) in °Brix, moisture content (dry basis) and Concentration of Phenolic Compounds (CPC) in mg of Gallic acid per 100 g of must.

In the Fig. 5 are shown the values obtained the soluble solids concentration and phenolic compounds concentration, another important feature in terms of differentiation of must. This results indicated the existing potential for applications of partial dehydration the grapes.

The mean test identified statistically significant modifications for the adaptation of must for winemaking purposes, purposes. It determined that the treatment with 22.9 °C and air velocity of 1.79 ms^{-1} was the most appropriate, which presented the highest value for the amount of total soluble solids, followed by the second best result for concentration of phenolic compounds. The other attributes are part of a range of value that does not interfere with the final destination of the must.

Another research work analyzed the physical-chemical changes concentration of total soluble solids (TSS) and phenolic compounds (CPC) after partial dehydration

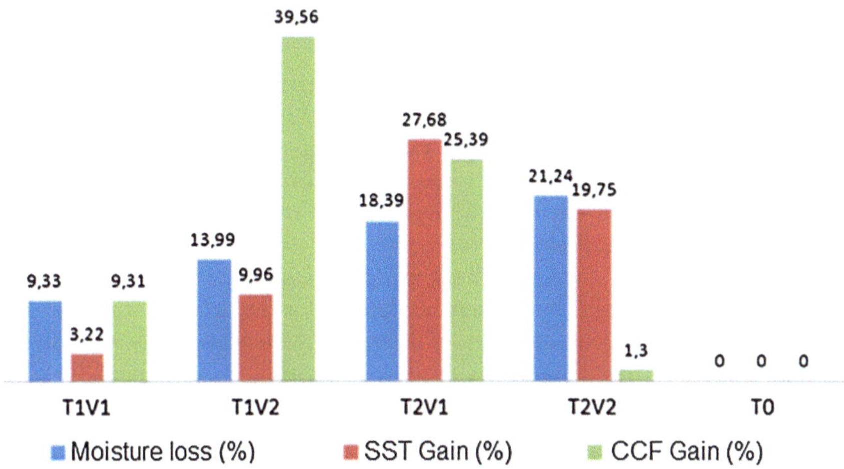

Fig. 5 Gain ratio by TSS and CPC through partial dehydration of clusters [26]

of Syrah grapes subjected to treatments combining two temperatures and one air velocity (T1 = 22.9 °C/1.79 ms^{-1} and T2 = 37.1 °C/1.79 ms^{-1}) and relative humidity of 40 % [27].

The results confirm one of the premises already investigated, which demonstrates that the effects of controlling thermal and psychometric parameters of dehydration process provide beneficial changes in concentration of phenolic compounds and total soluble solids.

The water loss of the grapes was approximately 14 % and the drying process lasted between 34 and 68 h for treatments T1 and T2, respectively. It was experimentally and statistically verified that the treatments promoted significant increase in TSS and CPC; however, for PC at temperature of 37.1 °C, the increase accounted was approximately 27 %, whereas, for temperature of 22.9 °C, it was only 12 %. For TSS, the increase was on average 12.47 ± 0.9 % between both treatments.

In the evaluation of the effectiveness of applied heat treatment, it is noted that both conditions are promising and could be applied if carried out under controlled conditions, once the smaller average gains for the main physicochemical parameters evaluated were 12.04 and 12.32 % to phenolic compounds and total soluble solids, respectively.

The results demonstrate that it is possible to moderately dry grapes, which consequently results in improvements in their chemical composition and can improve the quality of wine.

Other studies present in instrumentation and grape dehydration process control is to use spectral imaging for soluble solids prediction during drying. The Image analysis techniques can become an alternative to improve the process, reduce costs and make the process safer, ensuring quality food to consumers. Specifically spectral images show its potential use as a base for the system development detection and evaluation of chemical components, physical components and other properties [34–38].

With aim to correlate data from soluble solids of fresh grapes cv. Italy underwent partial dehydration (drying in an oven for 24 hours at 60 °C) with evaluating the process by spectral imaging (wavelength 480–670 nm, shown in Table 1). With aim to correlate data of soluble solids in grapes cv. Italy, fresh and subjected to partial dehydration (dried in an oven for 24 h at 60 °C) through spectral images, for wavelength 480–670 nm (Table 1).

The correlation was conducted by a principal component regression (PCR), with selection of the main components accounting for the data majority and with different soluble solids concentrations, to determine the model quality and validate the obtained data.

The average analyzed spectra showed a peak in the range 480–560 nm (Fig. 6). The spectral multivariate calibration to model showed prediction consistent results in the peak on spectral region (but does not consider the spectral regions where there was little or none difference between fresh and dry spectra) (Table 2).

The results show the application possibility a system with more precision for to automation and control grape systems dehydration, which are the most robust control and presenting a physic-chemical process control nondestructively and without interruption.

Table 1 Characterization of studied (Std—Standard deviation)

Treatment	Soluble solids (°Brix)	Std	Moisture content (wb)	Std
Fresh	13.47	1.27	76.06	8.32
Dehydrated grapes	19.14	1.63	52.73	12.69

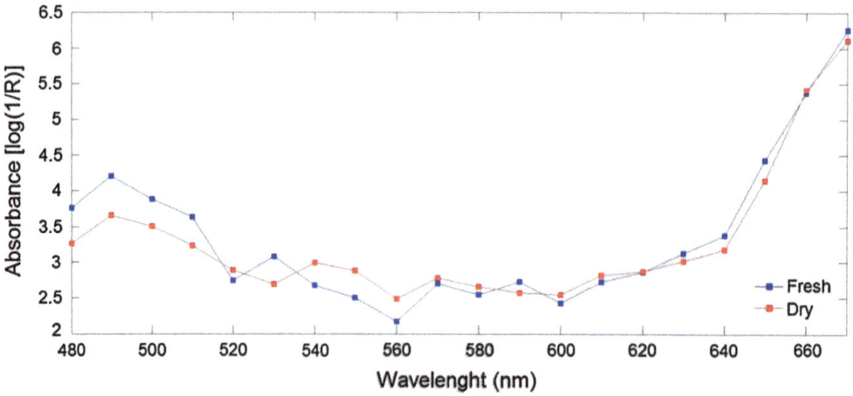

Fig. 6 Average spectra obtained before image process with spectral images in grapes partial dried and fresh

Table 2 Prediction model cross validation (Press—predictive residual erro sum of squares; Rmsecv—root mean square error of cross-validation)

Wavelength (nm)	Soluble solids				Press	Rmsecv
	Dry		Fresh			
	Mean	Std	Mean	Std		
480	17.00	1.62	14.68	1.56	6.33E−11	2.97E−06
490	17.03	1.22	14.82	1.26	7.22E−11	2.21E−06
500	16.82	1.30	15.16	1.34	3.95E−10	2.55E−06
510	16.87	1.36	15.02	1.43	5.44E−11	1.94E−06
530	19.97	1.24	14.84	1.52	9.62E−12	1.71E−06
540	17.25	1.54	15.39	0.97	1.33E−11	2.32E−06
550	17.46	1.63	15.17	1.09	1.31E−10	1.5E−06
560	17.42	1.71	15.19	1.11	1.31E−11	2.44E−06

4 Conclusions

The techniques of instrumentation, automatic control and process monitoring in real time as well as the quality evaluation with nondestructive techniques, facilitate the decision-making with aim to obtain better conditions before processing of the fruits. The techniques and results presented in this chapter show a path for the transfer of technology for the Vitiviniculture sector.

Acknowledgments Thanks to FAPESP (Fundação de Amparo a Pesquisa do Estado de São Paulo) and CAPES (Coordenação de Aperfeiçoamento de Pessoal de Nível Superior, Brazil), for supporting.

References

1. Penna, N.G., Hecktheuer, L.H.R.: Vinho E Saúde: Uma Revisão. Revista Informa **16**(1–2) (2004)
2. Jackson, R.S.: Wine science: principles, practice, perception. Food Science and Technology International Series. Academic Press, San Diego (2008)
3. Wine Institute: Statistics. San Francisco, California, United State of America (2011)
4. Mello, L.M.R.: Vitivinicultura Brasileira: Panorama 2011. Embrapa Uva E Vinho, Bento Gonçalves (2012a)
5. Mello, L.M.R.: Atuação do Brasil no Mercado Vitivinícola Mundial – Panorama 2011. Embrapa Uva E Vinho, Bento Gonçalves (2012)
6. Abrasuco: Associação Brasileira das Indústrias de Sucos de Uva. Indústria de Sucos de Uva Consome 50 % da Produção de Uvas Comuns (2012)
7. Sampaio, S.M., Queiroz, M.R.: Influência do Processo de Secagem na Qualidade do Cogumelo Shiitake. Engenharia Agrícola **26**(2), 570–577 (2006)
8. Azeredo, H.M.C., Jardine, J.G.: Desidratação Osmótica de Abacaxi Aplicada à Tecnologia de Métodos Combinados. Revista Brasileira De Produtos Agroindustriais **8**(2), 153–162 (2006)

9. Dionello, R.G., Berbet, P.M., Molina, M.A.B., Pereira, R.C., Viana, A.P., Carlesso, V.O.: Secagem de Fatias de Abacaxi in Natura e Pré-desidratadas por Imersão-impregnação: Cinética e Avaliação de Modelos. Ciência e Tecnologia de Alimentos **29**(1), 232–240 (2009)
10. Bradamente, S., Barenghi. L., Villa, A.: Cardiovascular Protective Effects of Resveratrol. Cardiovasc. Drug Rev. **22**(3), (2004)
11. Freitas, A.M., Detoni, A,M., Clemente, E., Oliveira, C.C.: Determinação de Resveratrol e Características Químicas em Sucos de Uvas Produzidas em Sistemas Orgânico e Convencional. Revista Ceres **57**(1), 1–5 (2010)
12. Lasa, M.J., Churruca, I., Simón, E., Arias, N., Milagro, F., Martínez, J.A., Portillo, M.P.: The Combination of Resveratrol and Cla does Not Increase the Delipidating Effect of Each Molecule in 3t3-L1 Adipocytes. Nutr. Hosp. **26**(5), (2011)
13. Barnabé, D., Filho, W.G.V.: Recuperação de Etanol a Partir do Bagaço de Uva. Revista Energia na Agricultura **23**(4), 1–12 (2008)
14. Guerra, C.C, Zanus, M.C.: Uvas Viníferas para Processamento em Regiões de Clima Temperado. Embrapa - Sistema De Produção, 4 (2003)
15. Bellincontro, A., De Santis, D., Botondi, R., Villa, I., Mencarelli, F.: Different postharvest dehydration rates affect quality characteristics and volatile compounds of Malvasia, Trebbiano and Sangiovese grapes for wine production. J. Sci. Food Agric. **84**, 1791–1800 (2004)
16. Curvelo-Garcia, A.S.: Práticas Enológicas Internacionalmente Reconhecidas. Ciência e Técologia em Vitivinicultura. **20**, 105–130 (2005)
17. Moreno, J.B., Cerpa-Caldrón, F., Cohen, S.D., Fang, Y., Qian, M., Kennedy, J.A.: Effect of postharvest dehydration on the composition of pinot noir grapes (Vitis Vinifera L.) and wine. Food Chem. **109**, 755–762 (2008)
18. Barbanti, D., Mora, B., Ferrarini, R., Cipriani, M.: Effect of various thermo-hygrometric conditions on the withering kinetics of grapes used for the production of “Amarone” And “Recioto” wines. J. Food Eng. **85**(3), 350–358 (2008)
19. Serratosa, M.P., Marquez, A., Lopez-Toledano, A., Merida, J., Medina, M.: Compostos Fenólicos Obtidos de Uvas Secas em Câmara com Temperatura Controlada da Casta Pedro Ximénez. Simpósio de Vitivinicultura do Alentejo. **8**, 123–131 (2010)
20. Código Internacional De Práticas Enólogas (CIPE). Paris, 2006
21. Lerma, N.L., Bellincontro, A., Mencarelli, F., Moreno, J., Peinado, R.A.: Use of electronic nose, validated by Gc–Ms, to establish the optimum off-vine dehydration time of wine grapes. Food Chem. **130**, 447–452 (2012)
22. Rolle, L., Giordano, M., Giacosa, S., Vincenzi, S., Segade, S.R., Torchio, F., Perrone, B., Gerbi, V.C.: Parameters of white dehydrated grapes as quality markers according to chemical composition, volatile profile and mechanical properties. Anal. Chim. Acta **732**, 105–113 (2012)
23. Mencarelli, F. And Tonutti, P.: Sweet, Reinforced and Fortified Wines: Grape Biochemistry, Technology and Vinification. Wiley, New York (2013)
24. El-Aouar, Â.A., Murr, F.E.X.: Estudo de Modelagem da Cinética de Desidratação Osmótica do Mamão Formosa (Caricapapaya L.). Ciência E Tecnologia De Alimentos **23**(1), 69–75 (2003)
25. Santiago, W.E., Tinini, R.C.R., Oliveira, R.A., Teruel, B.J.: Partial dehydration of ‘Niagara Rosada’ GRAPES (Vitis labrusca L.) targeting increased concentration of phenolic compounds and soluble solids. Afr. J. Biotechnol. **12**(46), 6474–6479 (2013)
26. Santiago, W.E., Silva, J.T.R., Teruel, B.J., Oliveira, R.A.: Adaptation of Niagara Rosada grape must to winemaking by partial cluster dehydration. Engenharia Agrícola **34**, 86–92 (2014)
27. Santiago, W.E., Teruel, B.J., Tinini, R.C.R.: Postharvest dehydration of ‘Syrah’ grapes (Vitis vinifera l.) under controlled temperature conditions with real-time monitoring of mass loss. Afr. J. Agric. Res. **10**, 229–234 (2015)
28. Constantini, V., Bellincontro, A., De Santis, D., Botondi, R.: And Mencarelli, F.: Metabolic changes of malvasia grapes for wine production during postharvest drying. J. Agric. Food Chem. **54**, 3334–3340 (2006)

29. Silva, J.C.T.R., Teruel, B.J.: Control system for forced-air cooling of horticultural products. Engenharia Agrícola. **31**, 621–630 (2011)
30. http://www.guildsomm.com/TC/stay_current/features/b/stamp/archive/2012/02/21/italy-39-s-adriatic-coast-wines-of-veneto
31. Serratosa, M.P., Marquez, A., Moyano, L., Zea, L., Merida, J.: Chemical and morphological characterization of Chardonnay and Gewürztraminer grapes and changes during chamber-drying under controlled conditions. Food Chem. **159**, 128–136 (2014)
32. Panceri, C.P., Gomes, T.M., DeGois, J.S., Borges, D.L.G., Bordignon-Luiz, M.T.: Effect of dehydration process on mineral content, phenolic compounds and antioxidant activity of Cabernet Sauvignon and Merlot grapes. Food Res. Int. **54**, 1343–1350 (2013)
33. Panceri, C.P., DeGois, J.S., Borges, D.L.G., Bordignon-Luiz, M.T.: Effect of grape dehydration under controlled conditions on chemical composition and sensory characteristics of Cabernet Sauvignon and Merlot wines. Food Sci. Technol. 1–8 (2015)
34. Baiano, A., Terracone, C., Peri, G., Romaniello, R.: Application of hyperspectral imaging for prediction of physico-chemical and sensory characteristics of table grapes. Comput. Electron. Agric. **87**, 142–151 (2012)
35. Baranowski, P., Mazurek, W., Wozniak, J., Majewska, U.: Detection of early bruises in apples using hyperspectral data and thermal imaging. J. Food Eng. **110**, 345–355 (2012)
36. Qiang, L., Mingjie, T.: Detection of hidden bruise on kiwi fruit using hyperspectral imaging and parallelepiped classification. Proc. Environ. Sci. **12**, 1172–1179 (2012)
37. Rajkumar, P., Wang, N., El Masry, G., Raghavan, G.S.V., Gariepy, Y.: Studies on banana fruit quality and maturity stages using hyperspectral imaging. J. Food Eng. **108**, 194–200 (2012)
38. Mishra, P., Herrero-Langreo, A., Barreiro, P., Roger, J.M., Diezma, B., Gorretta, N., Lleó, L.: Detection and quantification of peanut traces in wheat flour by NIR hyperspectral imaging spectroscopy using principal component analysis. J. Near Infrared Spectrosc. **23**(1), 15–22 (2015)

Production of Energy—The Second Generation Ethanol and Prospects

D.A.S. Leão, M.M. Conceição, L.S. Conrado, C.R.S. Morais, A.G. Souza, C.S.S. Lima, J.M. Silva Neto and F.L.H. Silva

Abstract Bioethanol is mainly produced from food sources, leading to criticisms regarding competition between food and fuels. The study of alternatives raw materials is very important. The reactions of hydrolysis of sisal bagasse were

D.A.S. Leão · C.S.S. Lima · J.M. Silva Neto
Center of Science and Technology, Federal University of Campina Grande (UFCG), Campina Grande, PB 58429-900, Brazil
e-mail: douglasasl@ufcg.edu.br

C.S.S. Lima
e-mail: sidneycssl@gmail.com

J.M. Silva Neto
e-mail: neto-silva@hotmail.com

M.M. Conceição (✉)
CTDR, IDEP, Federal University of Paraiba (UFPB), Av. Dos Escoteiros, S/N. Mangabeira, João Pessoa, PB 58055-000, Brazil
e-mail: martamaria8@yahoo.com

L.S. Conrado
Department of Chemical Engineering, Federal University of Campina Grande (UFCG), Campina Grande, PB 58429-900, Brazil
e-mail: libiaconrado@yahoo.com.br

C.R.S. Morais
Department of Materials Engineering, Federal University of Campina Grande (UFCG), Campina Grande, PB 58429-900, Brazil
e-mail: crislenemorais@yahoo.com.br

A.G. Souza
CCEN, IDEP, Federal University of Paraiba (UFPB), Av. Dos Escoteiros, S/N. Mangabeira, João Pessoa, PB 58055-000, Brazil
e-mail: gouveia@quimica.ufpb.br

F.L.H. Silva
Center of Technology, Federal University of Paraiba (UFPB), Av. Dos Escoteiros, S/N. Mangabeira, João Pessoa, PB 58055-000, Brazil
e-mail: flavioluizh@yahoo.com.br

J.M.P.Q. Delgado and A.G. Barbosa de Lima (eds.),
Drying and Energy Technologies, Advanced Structured Materials 63,
DOI 10.1007/978-3-319-19767-8_9

carried out in stainless steel reactor of high pressure. Sugar contents were determined by HPLC. The pretreatment was realized in the highest temperature and lowest acid concentration to obtain the maximum xylose concentration of 24,000 mg/L and maximum glucose concentration of 6700 mg/L. After the pre-hydrolysis the percentage of cellulose in the sisal bagasse concentrated in 107 % in function of hemicellulose reduction (30 %) and leaching of ashes (40 %). In the hydrolysis, the higher formation of glucose occurred in the highest temperature and highest acid concentration to obtain values of 4000 mg/L. The sisal bagasse has a high quantity of pentose from the hemicellulose. The hydrolyzate liquor can be used for the production of ethanol.

Keywords Biofuel · Alcohol · Sisal · Experimental

1 Introduction

The new technologies of ethanol of second generation permit to use agroindustrial residues to produce bioethanol, which means converting the celluloses from residues into fermentable sugars through acid or enzymatic hydrolysis. In Brazil the bioethanol from lignocelluloses presents excellent perspectives [1].

Different types of pretreatment method are used. These include pretreatment with dilute acid, steam pretreatment, wet oxidation, ammonia fibre explosion and alkaline pretreatment [2].

Brazil is the biggest producer of sisal (*Agave sisalana*) of the world; in despite of the importance of sisal to the country, there is still a lack for publications that treat the subject in a systematic way. According to data of National Company of Supply, the Brazilian production of sisal fiber in 2004 was 139.7 thousand tons, representing the percentage of only 4 % of the leaf. On the other hand, the solid residues from the defibration obtained for this amount of fiber, corresponding to 14 % of the leaf, were approximately 489.0 thousand tons. These residues are constituted of vegetal sap, particles of crushed parenchymal tissue and pieces of leaves and fiber of different sizes. In Brazil these residues are also named bagasse, and most of times they are wasted in the field [3].

Sisal presents low cost, high availability in Brazil, not used as source of food and high quantity of cellulose.

1.1 Lignocellulosic Biomass

Lignocellulosic materials consist mainly of three polymers: cellulose, hemicellulose and lignin. These polymers are associated with each other in a hetero-matrix to different degrees and varying relative composition depending on the type, species and even source of the biomass [4].

Cellulose is the main constituent of plant cell wall conferring structural support and is also present in bacteria, fungi, and algae. When existing as unbranched, homopolymer, cellulose is a polymer of ß-D-glucopyranose moieties linked via ß-(1,4) glycosidic bonds with well documented polymorphs. The degree of polymerization of cellulose chains in nature ranges from 10,000 glucopyranose units in wood to 15,000 in native cotton. The repeating unit of the cellulose chain is the disaccharide cellobiose as oppose to glucose in other glucan polymers [4–6].

Hemicellulose is the second most abundant polymer (20–50 % of Lignocellulose biomass) and differs from cellulose in that it is not chemically homogeneous. Hemicelluloses are branched, heterogenous polymers of pentoses (xylose, arabinose), hexoses (mannose, glucose, galactose) and acetylated sugars. They have lower molecular weight compared to cellulose and branches with short lateral chains that are easily hydrolysed [6, 7].

Lignin is the third most abundant polymer in nature. It is present in plant cell walls and confers a rigid, impermeable, resistance to microbial attack and oxidative stress. Lignin is an amorphous heteropolymer network of phenyl propane units (p-coumaryl, coniferyl and sinapyl alcohol) held together by different linkages [8, 9].

1.1.1 Sisal (*Agave sisalana*)

Among the few crops that suit the climate characteristics of the Brazilian semi-arid is the sisal (Fig. 1) that, as a semixerofila plant, it has been well adapted to the conditions of climate and soil, and it has the ability to withstand prolonged drought and high temperatures. It is responsible for maintaining of thousands of people in rural areas, and the economic foundation of some regions, such as the Sisal Territorial in Bahia [10, 11].

Brazil is the largest producer of this crop, and the state of Bahia accounts for 95 % of production and the Paraiba for 4.1 %, the remaining production is distributed between the states of Rio Grande do Norte and Ceará [11, 12].

The sisal is originally from Yucatan (Mexico) and its popular name comes from a native herb called zizal-xiu. Regarding the botanical name, sisal belongs to the monocot class, Liliflorea series, Agavaceae family, subfamily Agavoidea, gender Agave, and species Agave sisalana. The Agave name comes from the Greek agavos meaning admirable, magnificent; and sisalana means stiffness [13, 14].

In recent years it has been observed decline in its culture. Some of the factors contributed to this: the low rate plant utilization, therefore, only 4 % of harvested leaves are converted into salable product, the extracted natural fiber plant is used for the production of rope, twine, furniture, carpets, construction, which when compared to synthetic fibers it has low competitiveness, harm to farmers, and added to this, it is the non-use of waste from shredding [14].

Therefore, it is necessary, among other measures, the use of refining by-products, including the enhancement of fiber, to increase the sustainability of

Fig. 1 Sisal (*Agave sisalana*)

sisal activity and promote social inclusion of communities that remain of this culture [15]. The possible production of biofuel from the fiber a technological process highly favorable in this regard.

1.2 Second Generation Ethanol

The search for alternatives to ethanol production leads to search for new sources of biomass, in this context, it is worth remembering that Brazil is known for its large production potential renewable resources such as agricultural residues (such as bagasse from sugarcane sugar, rice straw, wheat straw to the oat hulls, and shell of sisal fiber fruits), forest residues (as wood chips, dust and remains of sawmill) and crops such willow and elephant grass [16].

Complete hydrolysis of the polysaccharide fractions of lignocellulosic material composition searches the full use of agro-industrial waste, increasing product yield in relation to the raw material.

Bioethanol, lignocellulosic ethanol or second generation ethanol is obtained through hydrolysis of polysaccharides present in the cell wall of biomass.

These polysaccharides are continually renewed in nature by the biotransformation of solar energy and CO_2 through photosynthesis [17]. The lignocellulosic biomass (grass, wheat straw, bagasse and corn stover, wood) is geographically distributed and its conversion into bioethanol helps to reduce emissions of greenhouse gases (GEF) in comparison with emissions from the combustion of gasoline [18–20].

The processes most commonly used in the conversion of biomass into fermentable sugars are enzymatic hydrolysis and acid hydrolysis. These processes invariably require a pretreatment step. From an environmental point of view, the enzymatic processes tend to further reduce emissions of the GEF, and produce bioethanol through cleaner production techniques [21].

As an example, it can mention that in the Brazilian context, where ethanol is produced using only a fraction of the raw material (sugarcane juice), the waste (surplus bagasse and straw) of composition lignocelluloseis obtained, which are subject to hydrolytic processes, providing sugars that can be fermented into ethanol [22]. In addition, this issue has now accompanied by the concept of "Biorefinery", which is similar to oil refineries, in design, however using biological material as opposed to fossil fuels for the production of transportation fuels, chemicals and energy. Industrial Biorefineries have been identified as the most promising route to the creation of a new industrial sector based on renewable raw materials. The national reality fits perfectly in this concept, thanks to the diversity of natural resources existing here and the country's vocation to these developments [23].

The production of ethanol from lignocellulosic biomass makes use of chemical/enzymatic pretreatments for the hydrolysis of cellulose and hemicellulose, providing carbohydrates (hexoses and pentoses) that can later be converted to ethanol by fermenting microorganisms [24].

Among the pre-treatments that have as main goal disrupting the lignocellulosic complex, acid pretreatment has been an interesting alternative, because in addition to disrupt the complex, causes hydrolysis of the hemicellulose, which results in a liquid fraction containing, in the case of sugarcane bagasse, mainly xylose. After the pretreatment, liquid containing the hemicellulosic hydrolysate is separated from a solid residue, mainly consisting of cellulose and lignin, called cellulignin [25].

The lignocellulosic raw material source contains 20–60 % cellulose; it can be fully converted to glucose by acid or enzymatic activity after the pretreatment step for the lignocellulosic complex clutter. In turn, glucose is characterized for being a monosaccharide used by most microorganisms, making this molecule an important building block for obtaining a vast range of substances of commercial interest, ranging from polymers to fuel [26].

In alcohol production using as raw material lignocellulosic biomass from agro-industrial residues and crops, it is employed three stages of processing which are: Pretreatment to separate cellulose and lignin; acid or enzymatic hydrolysis of cellulose to glucose conversion and ethanol fermentation by yeast [27].

1.3 Hydrolysis

The production of ethanol from lignocellulosic biomass makes use of chemical and enzyme pretreatments for the hydrolysis of cellulose and hemicellulose, providing carbohydrates (hexoses and pentoses) that can later be converted to ethanol by fermenting microorganisms [24].

Acid pretreatment is a process to break the rigid structure of lignocellulosic material in which hydronium ions breakdown and attack intermolecular and intramolecular bonds among cellulose, hemicellulose, and lignin in biomass hierarchy structure. The acid hydrolysis includes concentrated and dilute acid solutions where different levels of acid severity contribute to various biomass fractionated products [8].

Among the treatment is aimed at disrupting the lignocellulosic complex acid treatment has been an interesting alternative, because in addition to disrupt the complex causes hydrolysis of the hemicellulose, which results in a liquid fraction containing, in the case of sugarcane bagasse, mainly xylose. After the treatment, liquid containing the hemicellulosic hydrolyzate is separated from a solid residue, mainly consisting of cellulose and lignin, called cellulignin [25].

Acid hydrolysis has a major disadvantage where the sugars are converted to degradation products like tars. This degradation can be prevented by using enzymes favoring 100 % selective conversion of cellulose to glucose. When hydrolysis is catalyzed by such enzymes, the process is known as enzymatic hydrolysis [28, 29].

Enzymatic hydrolysis of natural lignocellulosic materials is a very slow process because cellulose hydrolysis is hindered by structural parameters of the substrate, such as lignin and hemicellulose content, surface area, and cellulose crystallinity [30].

Enzyme hydrolysis is usually conducted at mild conditions (pH 4.8) and temperature (318–323 K) and does not have a corrosion problem. The enzymatic hydrolysis has currently high yields (75–85 %) and improvements are still projected (85–95 %), as the research field is only a decade young [31].

Enzymatic hydrolysis is an environmentally friendly alternative that involves using carbohydrate degrading enzymes (cellulases and hemicellulases) to hydrolyze lignocelluloses into fermentable sugars [32].

1.3.1 Process for the Enzymatic Hydrolysis

Enzymatic hydrolysis causes the breaking of bonds in the cellulose structure β-1,4-glycosidic bonds in monomer units, β-glucose. This unit monomer is the one of interest for bioethanol production by process fermentation. The enzymes used in this process should produce a synergistic action, thus, it is necessary for a good conversion of cellulose to glucose the presence of enzymatic extract withthree major enzymes: endoglucanase, exoglucanase, and β-glucosidase.

The endoglucanase performs a random hydrolytic action in the amorphous regions of cellulolytic chain, reducing the degree of polymerization of the cellulose chain. The exoglucanase gradually attacks the ends of the cellulose, releasing celobioses subunits. These types of enzyme are not very active in the crystalline region but acts synergistically and cooperatively in the presence of endoglucanase in the crystalline region of cellulose chain. The β-glucosidases hydrolyze cellobiose and cello oligosaccharides produced by action of other enzymes.

For the enzymatic hydrolysis process of cellulosic materials to be efficient, it is necessary occurs an effective bond between the substrate and the enzyme, which in this case would be the cellulosic biomass and cellulolytic enzymes. So, the more exposed the cellulose is, the enzyme will act more active. Thus, the lignocellulosic material must be pretreated by physical and/or chemicals means. These pretreatments promote partial or total elimination of hemicellulose and lignin. These fibers produce non-fermentable compounds by yeast, commonly used for ethanol fermentation as Sacharomices. These fibers, act as barriers preventing that the spread of cellulolytic enzymes to the pulp occurs.

In enzymatic processing of lignocellulosic materials, enzymes soluble in buffer solutions are used and should be put in direct contact with the biomass. Overall the process is discontinuous under stirring, pH and constant temperature. The enzyme should be maintained at optimal pH and temperature for the majority of cellulolytic enzymes may act at full activity. Stirring must be adequate for the mass transfer of the enzyme to the solid surface of the lignocellulosic material with the goal that no resistance occurs. Such cares are essential to maintain the structural stability of the enzymes.

The ratio between the solution volume and the mass of cellulosic material as well as the enzyme load, which relates the "enzymatic activity" with the "total volume" of the solution and the mass of cellulosic material, should be controlled. Depending on the value of these parameters and their interactions the conversion of cellulose to glucose will be greater or lesser, consequently more or less the concentration of glucose in solution. It is necessary to optimize these parameters for enzymatic hydrolysates are less diluted.

2 Characterization of Sisal Bagasse for Production Ethanol

As an application sisal was characterized to obtain ethanol. The samples of sisal bagasse, variety *Agave sisalana*, were acquired in the rural zone of the Paraíba, Brazil.

The sisal bagasse was collected after the defibration. The samples were dried in the temperature of 75 °C into air circulation oven to constant mass [33] and grinded in knife mill for reduction of granulometry.

In the characterization of bagasse sisal the cellulose and hemicellulose were determined according to the methodology by Xu et al. [34]. Lignin content was determined using the Klason method with insoluble lignin in 75 % sulfuric acid, according to the methodology of Viera et al. [35]. Ash determination was made according to the methodology of Lutz [36].

Thermogravimetric curves (TG/DTG) of sisal bagasse show three steps of mass loss and ash content of 10.6 %. The first step, attributed to dehydration, occurred at 54–145 °C (4.7 % mass loss). The second step, attributed to decomposition of hemicellulose and cellulose, occurred in the interval of temperature of 145–384 °C (48.8 % mass loss). The third step, associated with cellulose and lignin decomposition, occurred in the interval of temperature of 384–681 °C (mass loss of 35.9 %). These thermogravimetric curves were obtained in Shimadzu Thermal Analyzer, synthetic air atmosphere, mass of 10 mg, heating rate of 10 °C/min and temperature of 30–700 °C.

Corradini et al. [37] in a study of thermal stability of cotton fibers, mention that the hemicellulose degraded between 240 and 310 °C, the cellulose in the temperature range from 310 to 360 °C, while the lignin between 200 and 550 °C. Also say that in many cases is not possible the different processes of degradation of the components of lignocellulosic fibers, due to the reactions are complex and overlap in the range from 220 to 360 °C.

Kim et al. [38] investigated the addition of lignocellulosic components to polymers given that the decomposition of cellulose occurs at temperatures below 350 °C by breaking the glycosidic bonds between carbons 1 and 4 of its monomers. They pointed out that above 400 °C is the breaking of C–O bonds of cellulose and volatile components such as CO and CH_4. On lignin, the researchers indicated higher thermal stability related to its high degree of condensation, leaving as residue, mainly phenolic compounds.

The diffractogram of sisal bagasse showed peaks representing the crystal planes characteristic of lignocellulosic materials: 15° (plane 101), 23° (plane 002) and 34° (plane 040). Primary components are lignin, cellulose and hemicellulose. Lignin and hemicellulose are amorphous macromolecules and cellulose molecules are randomly distributed, with a tendency to form intra- and intermolecular hydrogen bonds. These diffractograms were obtained in a Shimadzu XRD Diffractometer, CuKα radiation, 40 kV and 30 mA. Speed was 2°/min and scanning range of $2\theta = 10–50°$ [39].

2.1 *Prehydrolysis*

The hydrolysis and prehydrolysis reactions were made in a stainless steel reactor with capacity for 750 mL. The sulfuric acid was catalyzer. The time of reaction was 60 min, heating ramp of 15 min, ratio solid-liquid of 1:8 bagasse to acid [11].

The data of prehydrolysis (Table 1) indicated the influence of two input variables (temperature and acid concentration) over the output variables (glucose and xylose concentration).

Table 1 Experimental design matrix obtained in the prehydrolysis

EXP	Temperature (°C)	Concentration (%)	Glucose (mg/L)	Xylose (mg/L)
1	70	0.5	3042.42	11678.48
2	110	0.5	6734.36	23524.55
3	70	2.9	2798.62	12608.23
4	110	2.9	3728.81	17187.75
5	90	1.7	4388.78	16832.96
6	90	1.7	3480.50	16485.88
7	90	1.7	3748.78	16555.86

The experimental design 2^2 aimed the optimization of prehydrolysis of sisal bagasse [40]. The input variables were temperature and acid concentration, keeping the time constant in 60 min and the ratio solid to liquid 1:8 bagasse (7 experiments, 4 different experiments and 3 repetitions in the main point). The output variables were the concentrations of glucose, xylose, acetic acid, HMF and furfural. The real and codified levels have been chosen according to the adaptation of Lima [41], who studied the cashew peduncle prehydrolysis.

Equation 1 and 2 showed the linear regression models of experimental design 2^2 for xylose and glucose response. All the coefficients of models are statistically significant (95 % confident level).

$$\text{Xylose} = 16410.53 + 4106.40\,T - 1351.76\,C - 1816.64\,T.C \quad (1)$$

$$\text{Glucose} = 3988.90 + 1155.53\,T - 812.34\,C - 690.44\,T.C \quad (2)$$

where T—Temperature, C—concentration and T.C—temperature versus concentration (interaction)

The adaptation these models to the results obtained was verified by variance analysis (ANOVA). The Table 2 presents the responses of xylose and glucose in the prehydrolyzed liquor.

The regression models of experimental date of xylose and glucose concentration are statistically significant, 95 % confidence level, because the ratio F_{cal}/F_{tab} is higher than 1. The determination coefficients (R^2) are closer to 1.

In the prehydrolysis the models were statistically significant, then can be built the response surface for xylose and glucose. Figures 2 and 3 have showed the response surface of temperature and acid concentration for the xylose and glucose, respectively, to observe optimal range of process. In the Fig. 2 verified that the level +1 of temperature (110 °C) and the level −1 of acid concentration (0.5 %) indicated the maximum concentration of xylose (23,524 mg/L).

Lima et al. [42] studied the pretreatment of bagasse cashew stalk and obtained 5458.70 mg/L of xylitol in the pre-hydrolyzate liquor.

Table 2 ANOVA for the responses of xylose and glucose in the prehydrolysis

Statistical parameters	Xylose	Glucose
R^2	0.996	0.951
F test	30.72	2.10

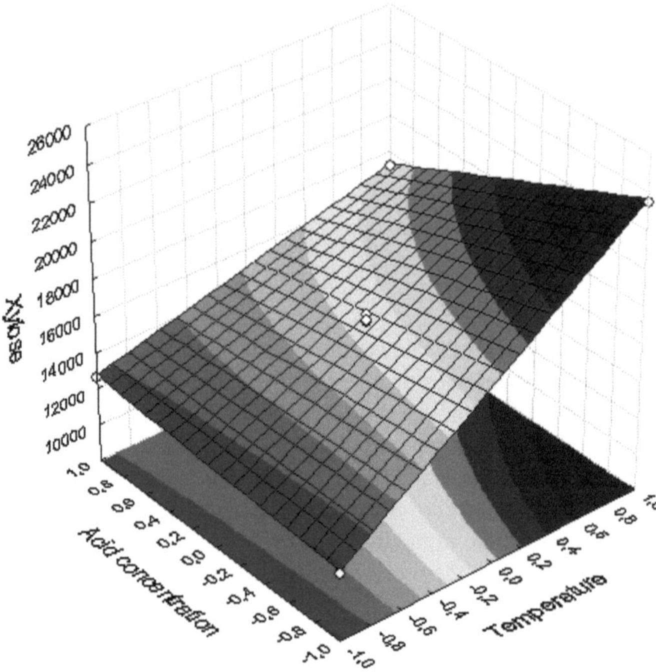

Fig. 2 Response surface for xylose concentration in the process of prehydrolysis

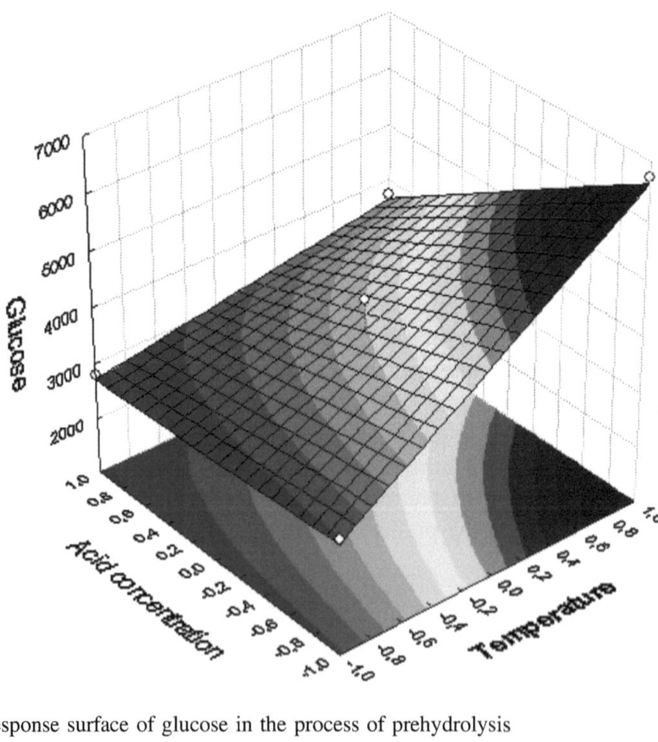

Fig. 3 Response surface of glucose in the process of prehydrolysis

Table 3 Sisal bagasse components in nature and pretreated

Component	Bagasse in nature (%)	Bagasse pretreated (%)
Cellulose	28.5	58.9
Hemicellulose	12.0	8.5
Lignin	26.0	26.0
Ash	10.0	6.1

Figure 3 represents the response surface of glucose concentration in the liquor of prehydrolysis, the highest value of glucose concentration was close to 6700 mg/L. The glucose presented the same behavior that the xylose. In the higher temperature (110 °C) and lowest acid concentration (0.5 %) optimized the xylose and glucose concentration.

The bagasse of this pretreatment has been washed and dried to 60 °C up to constant mass, then was analyzed for comparison between the bagasse before and after the pretreatment. Table 3 shows the values of cellulose, hemicellulose, lignin and ash, before and after the prehydrolysis.

After the prehydrolysis the percentage of cellulose in the bagasse of sisal is concentrated in about 107 % in function of the hemicellulose reduction of about 30 % and leaching of ashes of 40 %. The pretreatment has not affect lignin concentration.

2.2 *Hydrolysis of Sisal Bagasse*

The hydrolysis of sisal bagasse had as objective determine the conditions of higher liberation of glucose, verifying the influence of two input variables (temperature and acid concentration) over the output variables (glucose, xylose, acetic acid, HMF and furfural) to define the ranges that optimize the glucose answers and minimize inhibitors (Table 4).

Contents of sugars and aldehydes (HMF and Furfural) were established by High Performance Liquid Chromotography (HPLC), with a ProStar 210 (Varian) pump;

Table 4 Experimental design matrix with input and output variables obtained in the hydrolysis

EXP.	T (°C)	C (%)	Glucose (mg/L)	Xylose (mg/L)	Acetic acid (mg/L)	HMF (mg/L)	Furfural (mg/L)
1	120	1	207.58	767.69	432.47	4.54	83.97
2	120	5	104.99	3050.13	596.27	9.72	95.33
3	180	1	3208.56	11550.32	1267.32	193.95	617.84
4	180	5	3936.1	7593.59	129.3	372.45	463.98
5	150	3	1839.43	4255.45	966.77	39.78	470.51
6	150	3	2447.52	3587.67	139.99	33.69	54794
7	150	3	1935.96	3885.54	77.17	45.87	393.08

Table 5 ANOVA of response of glucose, xylose, acetic acid, HMF and furfural

Statistical parameters	Glucose	Xylose	Acetic acid	HMF	Furfural
R^2	0.976	0.920	0.549	0.828	0.801
F test	1.48	0.41	0.043	0.173	0.145

Manual injector with a 20 μL loop; ProStar 356 (Varian) refractive index detector and 284 nm UV/visible (aldehydes); Hi-Plex H (300 mm × 7.7 mm) stainless steel analytical column.

Operating conditions were as follows: Column temperature 40 °C; Mobile phase: MilliQ water, flow of 0.6 mL/min; Analysis time: 15 min and 60 min for sugar and aldehyde contents, respectively. Internal standard sugar solutions: glucose, xylose and arabinose, congeners 5-hydroxymethylfurfural—HMF (Aldrich 99.98 %) and furfural (Vetec 99.9 UV/HPLC) were used to quantify liquor components.

The adaptation of models to the results obtained was verified by variance analysis (ANOVA). The Table 5 presents the responses of glucose, xylose, acetic acid, HMF and furfural concentration in the hydrolyzed liquor.

The regression model of experimental date of glucose concentration are statistically significant, 95 % confidence level, because the ratio F_{cal}/F_{tab} is higher than 1. The Eq. 3 presented the 1st order model that represents the behavior of experimental data, due the determination coefficient (R^2) was 97.6 % for output variables of glucose.

$$\text{Glucose} = 1954.306 + 1708.023\,\text{T} \tag{3}$$

The coefficient presented in the regression was statistically significant with 95 % confidence level and T represents the temperature.

The analysis of variance was significant for the glucose with ratio F_{cal}/F_{tab} higher than 1, then can be built the response surface for glucose. The optimal formation of glucose occurred in the higher temperature and higher acid concentration (180 °C and 5 % acid concentration), obtaining values of glucose of 3936.1 mg/L.

Even with the temperature and acid concentration at higher levels than the pre-treatment, it is found that values of concentration in hydrolyzate liquor of glucose and xylose were lower. Thus, the hydrolyzate liquor can be used for the production of ethanol and xylitol.

3 Conclusions

Biofuels can contribute to regional development and reduce the emission of greenhouse gases. Moreover, it is very important to develop technologies to increase the production of ethanol, without competing with the agricultural areas for the food.

The residues of sisal defibration (sisal bagasse) have a high quantity of pentose from the hemicellulose. The conditions of optimal extraction of xylose in the prehydrolysis were in highest temperatures and lowest acid concentration, and in the process of hydrolysis the regions with the optimal extraction of glucose were in the highest temperatures and highest acid concentration. The hydrolyzate liquor of sisal bagasse can be used for ethanol production.

Sisal is a promising plant to production of ethanol. Both the cellulose and hemicellulose can be converted into ethanol by hydrolysis process.

Acknowledgments The authors would like to express their thanks to CNPq (Conselho Nacional de Desenvolvimento Científico e Tecnológico, Brazil) and CAPES (Coordenação de Aperfeiçoamento de Pessoal de Nível Superior, Brazil) for supporting this work

References

1. Soccol, C.R., Vandenberghe, L.P.S., Medeiros, A.B.P., Karp, S.G., Buckeridge, M., Ramos, L.P., Pitarelo, A.P.: Bioethanol from lignocelluloses: status. Bioresour. Technol. **101**, 4820–4825 (2010)
2. Bondesson, P.M., Mats, G., Guido, Z.: Ethanol and biogas production after steam pretreatment of corn stover with or without sulphuric acid. Biotechnol. Biofuels **6**, 11–16 (2013)
3. Andrade, R., Ornelas, J., Brandão, W: A situação atual do sisal na Bahia e suas novas possibilidades. Salvador, Rev. Bahia Agrícola. **9**, 1 (2011)
4. Agbor, V.B., Cicek, N., Sparlin, R., Berlin, A., Leven, D.B.: Biomass pretreatment. Biotechnol. Adv. **29**(6), 675–685 (2011)
5. Desvaux, M.: Clostridium cellulolyticum: model organism of mesophilic cellulolytic clostridia. FEMS Microbiol. Rev. **29**, 741–764 (2005)
6. Fengel, D., Wegener, G.: Wood Chemistry, Ultrastructure Reactions. Walter de Gruyter, New York (1984)
7. Saha, B.C.: Hemicellulose bioconversion. J. Ind. Microbiol. Biotechnol. **30**, 279–291 (2003)
8. Hendricks, A.T., Zeeman, G.: Pretreatments to enhance the digestibility of lignocellulosic biomass. Bioresour. Technol. **100**, 10–18 (2009)
9. Wei, J.H., Song, Y.R.: Recent advances in study of lignin biosynthesis and manipulation. J. Integr. Plant Biol. **43**(8), 771–779 (2001)
10. Da Silva, G.F.: Estratégias para inserção do território do sisal no programa de biodiesel. In: Bahia Análise & Dados. pp. 671–686. SEI, Salvador, BA (2009)
11. Lima, C.S.S., Conceição, M.M., Silva, F.L.H., Lima, E.E., Conrado, L.S., Leão, D.A.S.: Characterization of acid hydrolysis of sisal. Appl. Energy **102**, 254–259 (2013)
12. Gondim, T.M.S.: Caracterização de Frutos e Sementes de Sisal. Embrapa Algodão Campina Grande, Brazil, Circular técnica **127**, 1–6 (2009)
13. Silva, R.R.O.: Desenvolvimento da cultura do sisal nas regiões semi-áridas do nordeste brasileiro. Embrapa Cnpa, Campina Grande (1994)
14. Paiva, J.M.F., Frollini, E.: Sugarcane bagasse reinforced phenolic and lignophenolic composites. J. Appl. Polym. Sci. **83**, 880–888 (2002)
15. Brazil, Sistemas de Produção Cultivo do Sisal. Relatório técnico Embrapa Algodão. **5**, 1–12 (2006)
16. Galbe, M., Zacchi, G.: Produção de etanol a partir de materiais lignocelulósicos – Bioetanol de cana-de-açúcar, P e D para produtividade e sustentabilidade. Blucher, São Paulo (2010)
17. Lynd, L., Van Zyl, W., Mcbride, J., Laser, M.: Consolidated bioprocessing of cellulosic biomass: an update. Curr. Opin. Biotechnol. **16**, 577–587 (2005)

18. Slade, R., Bauen, A., Shah, N.: The greenhouse gas emissions performance of cellulosic ethanol supply chains in Europe. Biotechnol. Biofuels **2**, 15–19 (2009)
19. Rubin, E.: Genomics of cellulosic biofuels. Nature **454**, 841–846 (2008)
20. Doe, U.S.: Breaking the biological barriers to cellulosic ethanol: a joint research agenda, DOE/SC 0095, US Department Energy Office of Science and Office of Energy Efficiency and Renewable Energy (2006). http://genomicscience.energy.gov/biofuels/. Accessed 01 Dec 2014
21. Shelley, S.: Renewable feedstocks: trading barrels for bushels. Chemical Engineering, Rockville, MD, **116**, 16-19 (2009)
22. Silva, N.L.C., Bastos, H.B., Betancur, G.J.V., Maeda, R.N., Pereira Jr, N.: Produção de etanol de segunda geração a partir de biomassas residuais da indústrais de celulose. In: XVII Simpósio Nacional de Bioprocessos, Natal, Brazil (2009)
23. NREL USA: Chemical Analysis and Testing Task Laboratory - Conceito de biorrefinaria. Nat. Renew. Energy Lab **12**, 216–220 (2013)
24. Lin, L., Meng, X., Liu, P., Hong, Y., Wu, G., Huang, X., Li, C., Dong, J., Xiao, L., Liu, Z.: Improved catalytic efficiency of Endo-β-1,4-glucanase from Bacillus subtilis BME-15 by directed evolution. Appl. Microbiol. Biotechnol. **82**, 671–679 (2009)
25. Shi, Q.Q., Sun, J., Yu, H.L., Li, C.X., Bao, J., Xu, J.H.: Catalytic performance of corn stover hydrolysis by a new isolate Penicillium sp. ECU0913 producing both cellulase and xylanase. Appl. Biochem. Biotechnol. **164**, 819–830 (2011)
26. Knauf, M., Moniruzzaman, M.: Lignocellulosic biomass processing: A perspective. Int. Sugar J. **106**, 147–150 (2004)
27. Rossell, C.E.V.: Conversion of lignocellulose bomass (bagasse and straw) from the sugar-alcohol industry into bioethanol. Industrial perspectives for bioethanol. In: Franco, T. (ed.) São Paulo (2006)
28. Balat, M.: Production of bioethanol from lignocellulosic materials. Energy Convers. Manag. **52**, 858–875 (2011)
29. Pike, P.W., Sengupta, D., Hertwig, T.A.: Integrating biomass feedstocks into chemical production complexes using new and existing processes. Minerals Processing Research Institute, Louisiana State University, Baton Rouge. Los Angeles (2008)
30. Pan, X., Gilkes, N., Saddler, J.N.: Effect of acetyl groups on enzymatic hydrolysis of cellulosic substrates. Holzforschung **60**, 398–401 (2006)
31. Hamelinck, C.N., Van Hooijdonk, G., Faaij, A.P.C.: Ethanol from lignocellulosic biomass: techno-economic performance in short-, middle- and long-term. Biomass Bioenergy **28**, 384–410 (2005)
32. Keshwani, D.R., Cheng, J.J.: Switchgrass for bioethanol and other value-added applications: a review. Bioresour. Technol. **100**, 1515–1523 (2009)
33. AOAC. Official methods of analysis of the Association of Official Analytical Chemists. In: Horwitz, W. (eds.), 16ª edn. Washington (1997)
34. Xu, F., Sun, J.X., Liu, C.F., Sun, R.C.: Comparative study of alkali- and acidic organic solvent- soluble hemicellulosic polysaccharides from sugarcane bagasse. Carbohydr. Res. **341**, 253–261 (2006)
35. Viera, R.G.P., Rodrigues Filho, G., Assunção, R.M.N., Meireles, C.S., Vieira, J.G., Oliveira, G.S.: Synthesis and characterization of methylcellulose from sugar cane bagasse cellulose. Carbohydr. Polym. **67**(2), 182–189 (2007)
36. Lutz, A.: Métodos Físico-químicos para análises de alimentos. Instituto Adolfo Lutz. Edição IV, São Paulo (2008)
37. Corradini, E., Teixeira, E.M., Paladin, P.D., Agnelli, J.A., Silva, O.R.R.F., Mattoso, L.H.C.: J. Therm. Anal. Cal. **97**, 415–420 (2009)
38. Kim, H.S., Yang, H.S., Kim, H.J., Lee, B.J., Hwang, T.S.: Thermal properties of agro-flour-filed biodegradable polymer biocomposite. J. Therm. Anal. Cal. **81**, 299–306 (2005)
39. Martin, A.R., Mattoso, L.H.C., Martins, M.A., Silva, O.R.R.F.: Caracterização química e estrutural de fibra de sisal variedade Agave sisalana. Polímeros: Ciência e Tecnologia. **19**, 40–46 (2009)

40. Rodrigues, M.I., Iemma, A.F.: Planejamento de experimentos e otimização de processos, 2ª edn. Campinas, São Paulo (2009)
41. Lima, E.E.: Estudo das cinéticas de hidrólise ácida e fermentação alcoólica do bagaço do pedúnculo de caju para produção do álcool etílico. Doctorate Thesis, Process Engineering, Federal University of Campina Grande, Paraiba, Brazil, 2012 (In Portuguese)
42. Lima, F.C.S., Silva, F.L.H., Gomes, J.P., Silva Neto, J.M.: Chemical composition of the cashew apple bagasse and potential use for ethanol production. Adv. Chem. Eng. Sci. **2**, 519–523 (2012)

Effective Modelling of Phenomena in Over-Moisture Zone Existing in Porous Sand Mould Subjected to Thermal Shock

P. Popielarski and Z. Ignaszak

Abstract The problem concerns the thermo-physical properties of the porous ceramic material called mould to which the liquid metal is poured (in foundry industry). In the foundry processes the sand mould fulfils an auxiliary role only as technological tool, but its physical and technological properties determine the quality of the casting. The study includes the iron plate casting experiments poured in multi-component porous sand mould. The temperature fields in casting and in different zones of the mould were recorded. The determining of the thermo-physical properties of mould sand containing the over-moisture zone using simulation tests by Procast and NovaFlow & Solid foundry systems was the goal of this study. An originality of the related research is an attempt to take into account the effects of the global thermal phenomena occurring in the quartz sand bonded by bentonite-water binder, using the apparent (substitute) thermal coefficients. The majority of foundry simulation systems are not capable to modelling the phase transformation of water into vapour, vapour transport and its condensation in porous media (called green sand moulds). In these cases the application of heat and mass transfer substitute coefficients during modelling/simulation is an interesting effective way.

P. Popielarski (✉) · Z. Ignaszak
CAD/CAE Laboratory of Materials Technology, Poznan University of Technology, Poznan, Poland
e-mail: pawel.popielarski@put.poznan.pl

Z. Ignaszak
e-mail: zenon.ignaszak@put.poznan.pl

J.M.P.Q. Delgado and A.G. Barbosa de Lima (eds.),
Drying and Energy Technologies, Advanced Structured Materials 63,
DOI 10.1007/978-3-319-19767-8_10

1 Introduction—Specificity of High Temperature Use of Porous Materials in Foundry Applications

During heat transport through a multiphase porous material, its structure may be subject to modification resulting from phase changes of its components in the function of the value of the temperature and/or time. Phase changes may give rise to internal sources of latent heat. All of this affects dynamic changes of the temperature field. The intensity of these phenomena depends on the components, structure and porosity of the material, which are parameters affecting physical thermal and moisture fields.

In porous materials, moisture may occur as chemically bound water, sorptive moisture, water vapour or free water [1–3]. Of the most significant importance is water appearing on walls or inside pores and capillaries, bound with granular/fibrous material by means of van der Waals forces (sorptive moisture) or capillary forces.

Delving into theoretical notions, apart from molecular transfers, pores may also feature surface diffusion and capillary transport [1–8]. Simultaneously, the processes of sorption and desorption may occur accompanying the phenomena of the thermal disequilibrium of thermodiffusion. The occurrence and intensity of particular phenomena depend on the components, structure and porosity of the material and its properties affecting physical thermal and moisture fields.

The present work concerns a technological tool used in foundry which is a specific porous material containing a certain amount of water in intergranular pores.

The dynamic, thermally excited processes occurring in the material result from the conditions in which it is used. One needs to add that this tool plays only an auxiliary role in the process but has a significant impact on the final quality of the product, which is a casting of a particular alloy.

The porous ceramic materials described here are used in the manufacture of sand moulds widely used in foundries as disposable tools for casting different metal alloys. Operating as such a tool the sand mould fulfils a subsidiary yet important role with regard to the quality of the final casting.

The porous material used in our experimental tests is green sand where bentonite was used as a binder and cast iron as a casting material. The large majority of the world production of castings is made up of cast iron castings (74 million tonnes of castings in 2013, that is 72 % share in the world production) [9–11]. Out of that, about 50 % of cast iron castings are made in moulds from green bentonite sand. The physical properties of the mould material, to which the liquid cast iron is poured, have a decisive influence on the course of the physics–chemical metal-mould reactions and on the time dependent heat transfer rate from the casting especially during initial period of mould-casting thermal contact. The both determine the final casting quality. In the porous mould sand with the bentonite binder, where the amount of water added during the process does not exceed 4 %, is used to activate the purely physical mechanism of binding the matrix (quartz grains) using bentonite (type of clay). An important phenomenon accompanying the heating of the green sand mould by poured metal is the production of water vapour and then its condensation in the so-called over-moisture zone (local water content in over-moisture zone exceeds the value of initial process moisture content several times) in the deeper layers of the mould. This zone of moisture condensation, which is formed from the casting-mould interface, is characterized by low strength of mould sand and is the favourable factor to form the special casting defect (scab defect). This zone gets more space and moves into the mould to its external face. Simultaneously the water evaporates increasingly when the temperature of mould is close to 100 °C, the vapour is transported (presence of the pressure gradient), and further condenses in more and more distant (from casting) zones of the mould. The water vapour migration into the mould is accompanied by an additional heat transfer (parallel to conduction) related to a mass transfer (vapour). To sum up, the gradually dried and humidified zones of the mould, variable in time and space coordinates, are characterized by different local thermo-physical properties. The phenomena connected with the over-moisture zone affect obviously the dynamics of the heat transfer from the casting and the local time of solidification of its different wall sections. Dynamics of the heat transfer from the casting is of particular significance especially in the initial phase of the process of filling the mould cavity with the liquid cast iron, deciding on parameters of flow through the gating system, filling the mould cavity and distribution of temperature in the casting-mould system.

It is necessary to underline that there is a certain resemblance of the above described phenomena in a sand mould with the process of intensive drying of porous materials, e.g. in food, agricultural and ceramic industry. In such a case, the existence of the over-moisture zone (in which condensation arose from the fact of exceeding by the vapour air mixture of 100 % of relative humidity) is a transient stage and the drying process ends with the complete removal of water to the level of relative humidity corresponding to the ambient conditions.

In a casting mould, the one-sided effect of a hot casting and its drying is interrupted when the cooling casting reaches the temperature of 400–500 °C. In order to obtain a casting, the mould is destroyed (the process is known as knocking out). The acting of a mould on a casting is interrupted and the scattered fragments of the mould, with a varied amount of water, are subject to the recycling process (reclamation). Having replenished the used components, it is used in the construction of new moulds.

The aim of the study is to find the effective way to take into account in the foundry simulation system the presence of over-moisture zone and its impact on the cooling rate of the casting. In the simplified approach presented in the paper application of apparent (substitute) thermo-physical coefficients of the porous wet medium (moulding sand) in function of temperature in the classic foundry simulation system database (Procast and NovaFlow & Solid [12, 13]) were introduced.

2 Problem Definition—Different Phenomena in Porous Materials

2.1 Analysis of Mean Phenomena in over-Moisture Zone

The thermo-physical properties of the porous mould material to which the liquid metal is poured, have a decisive influence on the course of the physicochemical metal-mould reactions, and on the time dependent heat transfer rate from the casting, both determine the final casting quality. In the porous mould sand with bentonite binder there the technological water amount does not exceed 4 %, is used to activate the purely physical mechanism of binding the matrix (quartz grains) using bentonite (kind of clay). An important phenomenon accompanying the heating of the so-called green sand mould by poured metal is the production of water vapour and then its condensation in the over-moisture zone in the deeper layers of the mould. In the over-moisture zone local water content exceeds several times the value of initial technological moisture content. Studies conducted in [14, 15] according to the methodology shown there, allowed to trace the changes of humidity and water vapour pressure in the over-moisture zone since the first contact of metal (aluminium casting wall) with the mould. The humidity and temperature curves, recorded 3 mm from the casting wall, are presented in Fig. 1. In order to explain the distribution of humidity in a mould, research conducted by Szmigielski [14] used the conductometric method, which uses the change of electric

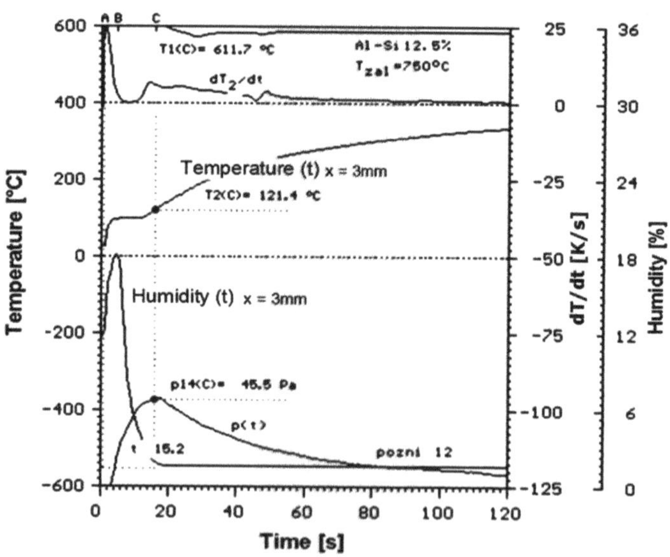

Fig. 1 Humidity and temperature curves recorded in over-moisture zone $x = 3$ mm (changes in pressure at this point and changes in temperature and the cooling rate of a casting were also presented) [14]

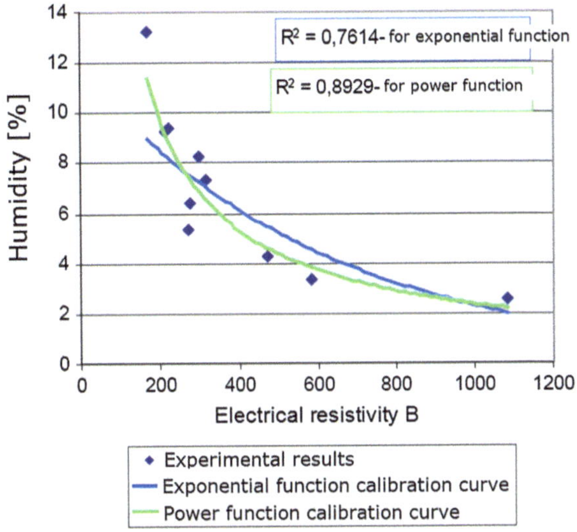

Fig. 2 Relationship between electrical resistivity B and humidity of the moulding sand [16]

conductivity of mass which changes along with the change of its humidity. This method allows to perform continuous measurements. Its disadvantage, however, is a relatively low degree of accuracy. Figure 2 shows the determined dependencies of electrical resistance B and humidity in moulding sand [16, 17]. As the measurements show, the identification of humidity is credible only in the range from 2 to 10 %.

This zone of moisture condensation, which is formed from the casting-mould interface is characterized by low strength of mould sand and is the favorable factor to form the special casting defect (scab defect) [14, 18]. This zone does getting more space and is moved into the mould to its external surface. Simultaneously the water evaporates increasingly when the temperature of mould is close to 100 °C, the vapour is transported (presence of pressure gradient), and further condenses in more and more distant (from casting) zones of the mould.

Forming and movement of the over-moisture zone is described in the publications [19, 20], where the authors distinguished three zones in the heated green sand: dry sand zone, transitory zone (evaporation, movement and condensation of the water vapour) and fresh (primary, external) sand zone. These zones are separated with the apparent surfaces of evaporation and condensation. On these surfaces, the characteristic temperatures correspond to phase transitions occurring in the mould space (only water boiling temperature is considered, this space is called over-moisture zone). The transient location of these surfaces are expressed as a simplified time function (1) and (2), (Fig. 3) [19].

$$x_1^2(t) = K_1 \cdot t \tag{1}$$

$$x_2^2(t) = K_2 \cdot t \tag{2}$$

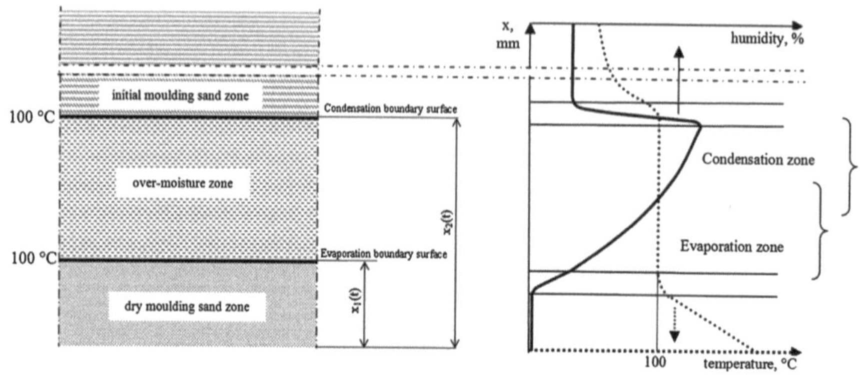

Fig. 3 Slice of the over-moisture zone and adjacent zones in the typical green sand mould subjected to thermal shock [19]

However, the authors analyse only a movement of the over-moisture zone (the boundary between the dry zone and the transition zone), omitting the humidity accumulation in the transition zone. The similar relations are proposed in the [21].

In the over-moisture zone, the partial filling of pores with water gives rise to a significant decrease in gas permeability and lowering the durability of moulding sand resulting from the increase in humidity of binders from bentonite paste [18].

The water vapour migration into the mould is associated with an additional heat flux and with a mass transfer (vapour) [22]. Resuming, the gradually dried and—over-humidified zones of the mould (in the deeper layers), variable in time and space coordinates, are characterized by differing local thermo-physical properties. The phenomena connected with over-moisture zone affect also on the dynamics of the heat transfer from the casting and on the local time of its wall solidification.

2.2 *Classic no-Foundry Heat and Moisture Transfer Modelling*

Forming and movement of the water vapour in the moist, porous bodies subjected to the thermal shock (contact with the liquid alloy) are a complex problem in terms of phenomena description using elementary models (Fig. 4).

The existing models can be formulated by expressing source of evaporation heat and water vapour penetration separately, which boils down to coupling of the fields: temperature (Fourier–Kirchhoff model) and humidity distribution (by intermediate using of Darcy and Fick models). Also, a substitute coefficient of heat transfer in the porous body can be introduced and the problem can be reduced to the temperature field modified by the presence of the humidity. The knowledge about amounts of the water vapour flowing through the mould (stream of mass) and transferring a defined amount of heat has become a base for balanced expression of the heat flux.

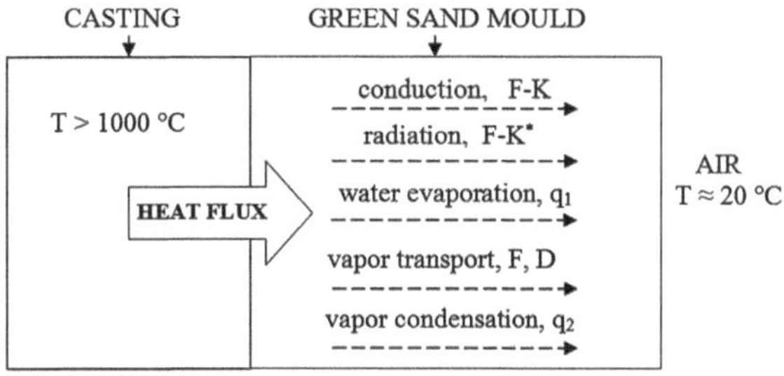

Fig. 4 Schematic diagram of the heat transfer mechanisms in casting-mould system described by models: *F-K* Fourier–Kirchhoff, *F-K** modified Fourier–Kirchhoff, *F* Fick, *D* Darcy, q_1 negative internal heat sources—evaporation, q_2 positive internal heat sources—condensation

Formally, the coupled model of phenomena of appearance and movement of the over-moisture zone should include phenomena of the energy flow in the presence of negative—evaporation—and positive—condensation—sources of heat (the Fourier–Kirchhoff equation), along with equation of movement (penetration) of fluid, i.e. water vapour and water in the porous medium, caused by the pressure difference (the Darcy equation).

The coupled heat and moisture transfer modelling is described in many publications but only for the cases of buildings walls, porous structural materials and fibrous insulation [23–26] where the temperature of process is always lower then 100 °C and its dynamics is significantly lower then in green sand mould during casting processes.

There is a resemblance between moulding sand and ceramic (mineral) building materials. The majority of available building materials, due to their capillary-porous make-up, make it possible to imbibe water which may be absorbed deep into the material filling entirely or partially its pores [25, 26]. Often an important factor is also the shape of pores (changing from narrow crevices into spherical bulbs) and the distribution of their dimensions and spatial arrangement in the material. Generally, for models describing the speed of fluid transfer in a porous structure and taking into account the mechanism of forced convection, a very important property is the permeability of the porous material. It is its capability to make it possible for a liquid, fluid or gas to permeate through porous space under the influence of the gradient of pressure. The difference in pressure conditioning the flow of liquid may be obtained through overpressure or under pressure in relation to atmospheric pressure. The vector of the speed of the filtration of liquid flowing in a porous medium u is directly proportional to the pressure gradient *grad P* and inversely proportional to the viscosity of the flowing liquid μ and the proportion coefficient, known as permeability K, is a constant parameter characteristic for a given porous medium:

$$u = -\frac{K}{\mu} gradP \tag{3}$$

The negative sign in the above vector equation comes from the fact that the vector of the speed of liquid filtration is adversely directed towards the vector of the pressure gradient.

Transfer of water in porous materials also depends on the size, shape and type of pores, and appearance in them of particular forces and mechanisms causing a flow to take place. In models based on the analysis of transfer processes in pores [3, 5, 26] two phenomena can be singled out, namely diffusion, as a notion describing the flow of water vapour and the so-called capillary conductivity, which describes the process of liquid water transfer in pores. In such a type of models, the energy-related aspect and coupling with the Fourier–Kirchhoff model are still not taken into account, that is the temperature field is not treated to have an essential impact on the phenomena of transfer. The result is the distribution of the field of the amount of humidity in the modelled space.

Rose considers in his model three transfer mechanisms, singling out, apart from the capillary stream, two types of diffusion: surface diffusion and water vapour diffusion. Another division proposed by Gertis and Werner [27], is the following: diffusion, laminar flow and Knudsen molecular flow.

Water vapour transport is carried out through diffusion, filtration and molecular movement, whereas water transfer happens as a result of the capillary phenomenon, diffusion and filtration caused by the pressure gradient. Separating these processes is very difficult as the participation of each of them depends on a variety of factors, including their components, structure in the material, construction solutions used, ambient conditions, etc. Searching for the model integrating heat and humidity transport was analysed in [28, 29]. The appearance of condensate causes a sharp local decrease in insulation, especially in the over-moisture zone and in its proximity.

In [29] the interesting model of coupled heat and moisture transfer with phase change and mobile condensates in fibrous insulation is presented. The new model considered the moisture movement induced by the partial water vapour pressure, a super saturation state in condensing region as well as the dynamic moisture absorption of porous materials and the movement of liquid condensates. This model needs the numerous data base parameters.

For numerical calculations, the authors [29] used a relatively complete physical model composed of three layers: thin inner fabric (0.1 mm) next to the human skin, thick porous fibrous batting (10 mm) and the other layer of fabric (0.1 mm) next to the cold environment.

Bearing in mind the transfer of heat and humidity through these layers (one-dimensional model 1D was adopted), on the basis of the balance of energy conservation [30–32], at the point of coordinate x and in time t the phenomena happening are described by the model equation for the temperature change (4). It contains elements representing: conduction of heat through the structure of porous material—λ/C $(\partial^2 T/\partial x^2)$, thermal radiation—$(\partial F/\partial x)$ and an element (internal source) connected with the phase change (vaporization, condensation) of the substance "w" appearing or disappearing in time—L_{vol} $(\partial w/\partial \tau)$.

$$C(x,t)\frac{\partial T}{\partial \tau} = \frac{\partial}{\partial x}\left(\lambda(x,t)\frac{\partial T}{\partial x}\right) + \frac{\partial F}{\partial x} + L_{\mathrm{vol}}(x,t)_{\text{phase chang}} \cdot \frac{\partial w}{\partial \tau} \tag{4}$$

Then, according to mass conservation, water vapour transfer in the inter-fiber void is controlled by the moisture transfer equation. The free water may diffuse when the free water is in liquid form and its content exceeds a critical value, also according to the mass conservation. The differential equations were solved by the Implicit Finite Difference Method, with assumed initial conditions simulating different practical circumstances. The distribution of water content and heat flux are calculated and compared with experiment. The calculated results are in general agreement with our experimental investigation. The model proposed in [29] pertains to coupled phenomena taking place in wet porous material in temperature not exceeding 40 °C. Its application to the case being solved in this study (wet porous sand mould subjected thermal shock) would require the development of new complex procedure and implementation of coupled models. Their inclusion, e.g. into models existing, e.g. in Procast system, would be a great challenge, yet of little practical value in foundry process modelling.

2.3 New Proposition—Effective Modelling Including Simplification from Experimental Point of View

The solutions proposed in [29] are used to calculate in time and the variability of the content of water and the stream of heat in function x and t. The vaporization effect, humidity diffusion, condensation and water movement in a moist porous material cause the stream of heat to be increased. These phenomena also occur in green sand. However, in casting practice these problems come down to only the issue of heat extraction from a casting. Heating and drying processes are of secondary significance in the casting processes modelling, also because of the fact that in order to solve the Eq. (4) it is necessary for their simulation to use thermophysical parameters (database) which are difficult to be determined individually in the above described complex reality of the moist porous material.

Essential elements for a casting include the thermal effects of these processes and the impact they have on the process of its solidification. With such an assumption, modelling phenomena occurring in moulding sand should allow for only energy-related influence on the casting and should be referred in this way to the specificity of the casting process, taking into account the auxiliary role of the mould in this process.

To sum up, a foundry mould is a multi-component ceramic composite material containing a granular and/or fibrous matrix and also other materials including a material called the binder. These components are often characterized by diverse fire-resistance and influence heat parameters in high temperature conditions, for example, more than 1000 °C, the space structure, internal/external porosity and chemical composition are significantly modified. Moreover, new phases occur as a result of endo- or exothermic transformations. Together with these transformations the mechanical and thermal properties are modified and that depends on the time (intensity) of heating. This process is significantly non-stationary [33, 34].

The testing and solution of thermal processes carried out in multi-component macro-porous materials can be conducted using a two-way method [34]:

1. The coupled model—formulation of heat and mass transfer equations solved by available methods, on condition that the thermal and diffusion characteristics are known as a function of unclear specific conditions and the structure type of material
2. The porous body is considered as quasi-homogenous and one thermal transfer equation is formulated (alternatively taking into consideration the internal heat sources, positive or negative). The material characteristics used in this model reflect all phenomena accompanying heat transfer (conduction, radiation, convection and also the mass transfer). These coefficients, called thermo-physical or substitute, require detailed analysis.

If a three-phase aggregation state as the set of solid particles with varied shape and size, separated by complex micro-spaces filled with liquid or/and gaseous phase, is recognised as generally being the case, then the successive fundamental mechanisms and phenomena will be responsible for heat transfer:

- conduction by each separated solid particle, creating a frame of porous body or performing the role of a binder or special component,
- conduction by the area of direct contact between the particles mentioned,
- molecular conduction in components of liquid existing in the pores,
- heat exchange between the solid particles and pore components,
- radiation between the solid particles,
- radiation through the semi- and transparent solid particles,
- convection in the fluids (liquid or/and gas), in the dispersed pores.

In principle, while solving the practical problems of heat transfer for porous materials it's not necessary to improve the physical nature of phenomena described e.g. considering the molecular-atomic structure.

The Fourier–Kirchhoff equation, which describes the temperature field and the porous space, can be considered as quasi-homogenous, and thermal conductivity coefficient expresses the effects of all heat transfer phenomena. It may be an obvious simplification but the form of Fourier–Kirchhoff equation (5), where the effective conductivity components are present, (conduction—λ_{cond}, convection —λ_{conv}, radiation—λ_{rad}), becomes more complicated (for 1-D dimension (5))

$$C(T)\frac{\partial T}{\partial \tau} = \frac{\partial}{\partial x}\left(\lambda_{\text{cond},x}(T)\frac{\partial T}{\partial x}\right) + \frac{\partial}{\partial x}\left(\lambda_{\text{conv},x}(T)\frac{\partial T}{\partial x}\right) + \frac{\partial}{\partial x}\left(\lambda_{\text{rad},x}(T)\frac{\partial T}{\partial x}\right) + \dot{q}_v \quad (5)$$

The application of a substitute coefficient of heat conduction in the case of wet moulding sand is insufficient. It is necessary to take into account mass and thermal effects of humidity transfer in the form of vapour and condensate. In research done in the past about acceptance by a mould, the problem of thermal effects of water transfer was not fully expressed in numerical models (some casting simulation systems take into account the emergence of over-moisture zone, however the model is not thoroughly explained and the obtained results are not in agreement with the experiment) [22]. The simplified approach proposed in our research [35], takes into account the said phenomena. The thermal model included the substitute coefficient of heat conduction λ_{subst} and the substitute thermal capacity of moulding sand C_{subst} as dry porous material (for initial content of water).

Next, two notions connected with phenomena taking places wet porous material were specified:

- modified substitute thermal capacity $C_{subst/modif//vapour}$ where it was expressed globally: C_{subst}—dry porous material and an element (internal heat source) connected with phase change (evaporation) of water "w" disappearing in time t by evaporation—$\rho \cdot L_{vapour}$ $(\partial w/\partial \tau)$, where ρ_{subst} is the average density of dry/wet moulding sand.

$$C_{subst}(T)\frac{\partial T}{\partial t} = \frac{\partial}{\partial x}\left(\lambda_{subst}(T)\frac{\partial T}{\partial x}\right) - \dot{q}_{vaporization} \tag{6}$$

$$C_{subst}(T)\frac{\partial T}{\partial t} = \frac{\partial}{\partial x}\left(\lambda_{subst}(T)\frac{\partial T}{\partial x}\right) - \rho \cdot L_{vaporization}(T)\frac{\partial T}{\partial t}\frac{\partial w_{vapor}}{\partial T} \tag{7}$$

$$C_{subst/modif/vapor}(T) = C_{subst}(T) + \rho \cdot L_{vaporization}(T)\frac{\partial w_{vapor}}{\partial T} \tag{8}$$

- the condensation latent heat (as direct internal heat source existing in the basic equation, proportional to total quantity of condensed water—[%]) with following parameters: condensation energy $L_{condensation}$, start temperature $T_{ini/condensation}$ and time period $t_{condensation}$ of condensation energy emission

$$C_{subst/modif/vapor}(T)\frac{\partial T}{\partial t} = \frac{\partial}{\partial x}\left(\lambda_{subst}(T)\frac{\partial T}{\partial x}\right) + \dot{q}_{condensation} \tag{9}$$

$$\dot{q}_{condensation} = L_{condensation}\int \frac{\partial w_{condensation}}{\partial t} = L_{condensation}\sum \frac{w}{t_{condensation}} \tag{10}$$

The average condensation latent heat intensity is as follow (for intensity not changing in time $t_{\text{condensation}}$):

$$\dot{q}_{\text{condensation}} = \rho \cdot L_{\text{condensation}} \frac{\int \frac{\partial w_{\text{condensation}}}{\partial t} \mathrm{d}t}{t_{\text{condensation}}} = \rho \cdot L_{\text{condensation}} \cdot \frac{w_{\text{condensation}}}{t_{\text{condensation}}} \tag{11}$$

The phenomena of moisture transfer are rarely included in modelling of foundry processes. Virtualization of casting process is used widely in foundry technological offices and allows to optimize technology concept (condition: experimental validation). Unfortunately in the foundry simulation systems, such as Procast [12] and NovaFlow & Solid [13], phenomena related to water vapour migration are not considered in the simulation models.

The Magmasoft [36] software allows to define the initial humidity of the mould sand and in the database, the value of the specific heat in a temperature range of approx. 100 °C is increased by value of the evaporation heat of water contained in the sand. The value of the considered evaporation heat is not related to the initial humidity, which determines the amount of the water in the mould sand. The simulations performed in the Magmasoft system allow formal representation of phenomena related to forming of the over-moisture zone, while the results obtained for the various initial humidity values are not compatible with intuitive prediction. The deviations of simulated times of solidification of cast iron plates from real times determined by experiment are approximately 50 % higher, which means that the energetic validation was not achieved. To forecast the casting quality by virtual means, the Computer Aided Engineering systems applied in foundry (e.g. Procast, Magmasoft, NovaFlow & Solid) are more effective, when the thermo-physical mould characteristics better correspond to reality with regard to the numerous simplifications of the mathematical model.

3 Methodology of Experimental and Simulation Studies

In order to receive verifiable results concerning the phenomena occurring in wet moulds, two experiments of casting cast iron into such moulds were performed. The first experiment (Exp. 1) was performed in the real conditions of a casting-mould system (plates thicknesses 16, 30 and 40 mm poured in the porous multi-component mould silica sand + active bentonite 7.0 % + coal dust 1.0 % + water 3.5 %). Arrangement of the castings in the mould is shown in Fig. 5. The cast iron and green sand parameters are shown in Table 1. The aim of the experiment was to measure and record the temperature variation in selected points of the casting and the sand mould. To conduct the experiment, special measuring cores needed to be prepared out of the studied green sand (Fig. 6). In the cores, the mineral insulated thermocouples of type K were placed, in the distances of 6, 15, 35 and 75 mm to the casting-mould interface and in the geometrical centre of each casting. The measurement results were recorded using the CASTOR 2—data acquisition system. In the second experiment (Exp. 2) plates 10, 20 and 30 mm thickness were cast

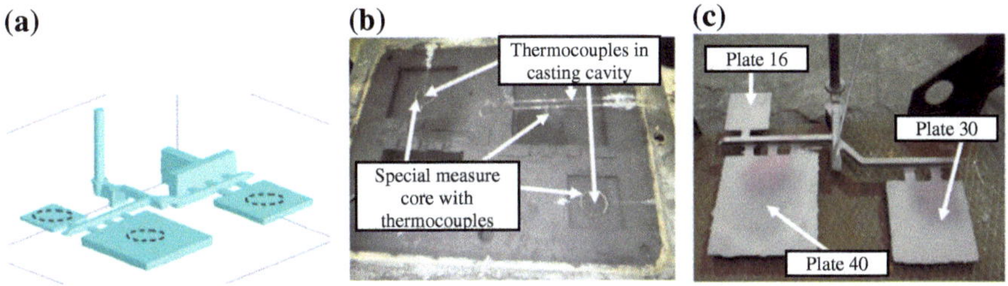

Fig. 5 First experiment (Exp. 1). Casting arrangement in the mould (800 mm × 600 mm × 250 mm). **a** CAD model with measurement core position, **b** mould instrumentation, **c** raw casting knocking out and blasted

Table 1 The cast iron and green sand parameters

<table>
<tr><td colspan="8">Cast iron—final composition</td></tr>
<tr><td>C</td><td>Si</td><td>Mn</td><td>P</td><td>S</td><td>Cr</td><td>Cu</td><td>V</td></tr>
<tr><td>3.51</td><td>1.99</td><td>0.53</td><td>0.18</td><td>0.09</td><td>0.13</td><td>0.39</td><td>0.02</td></tr>
<tr><td>Mo</td><td>Ni</td><td>Ti</td><td>Al</td><td>Co</td><td>Sn</td><td>Te</td><td>Mg</td></tr>
<tr><td>0.01</td><td>0.06</td><td>0.02</td><td>0.004</td><td>0.01</td><td>0.008</td><td>0</td><td>0</td></tr>
<tr><td colspan="8">Fusion and pouring parameters</td></tr>
<tr><td colspan="3">Furnace charge: pig iron 30 %, steel scrap 35 %, recycle scrap 35 %</td><td colspan="3">Superheating temperature in the furnace $T_{sh} = 1505$ °C</td><td colspan="2">Tapping temperature $T_{tapp} = 1450$ °C</td></tr>
<tr><td colspan="3">Ladle capacity (over lip)—1500 kg</td><td colspan="3">Pouring temperature $T_p = 1345$ °C</td><td colspan="2">Pouring time $T_z = 11$ s</td></tr>
<tr><td colspan="8">Green bentonite-sand parameters</td></tr>
<tr><td colspan="6">Dry components: silica sand 92 %, active bentonite 7.0 %, coal dust 1.0 % + water 3.5 %</td><td colspan="2">Density (wet state): $\rho = 1540$ kg/m^3</td></tr>
<tr><td colspan="8">Mechanical/technological parameters: compression strength 0.24 MPa, permeability 195×10^{-8} m^2/Pa/s, humidity 3.5 %, compatibility 40 %, calcination losses 3.55 %</td></tr>
</table>

(Fig. 7). Temperature was recorded in the geometric centre of each casting. The mould was made using green sand of a lower moisture level. The measuring core was used only for the plate 30 mm thick. The more precise assembly of the measuring core was attained by the application of the lost foam method (model made out of expanded polystyrene burnt during casting). This method does not require the mould to be disassembled in order to remove the model and assemble the measuring core. In order to determine the influence of the initial humidity of moulding sand on the creation and movement of the over-moisture zone, humidity was diversified—the second experiment used moulding sand of lower humidity (2.5 %) than in experiment Exp. 1 (3.5 %).

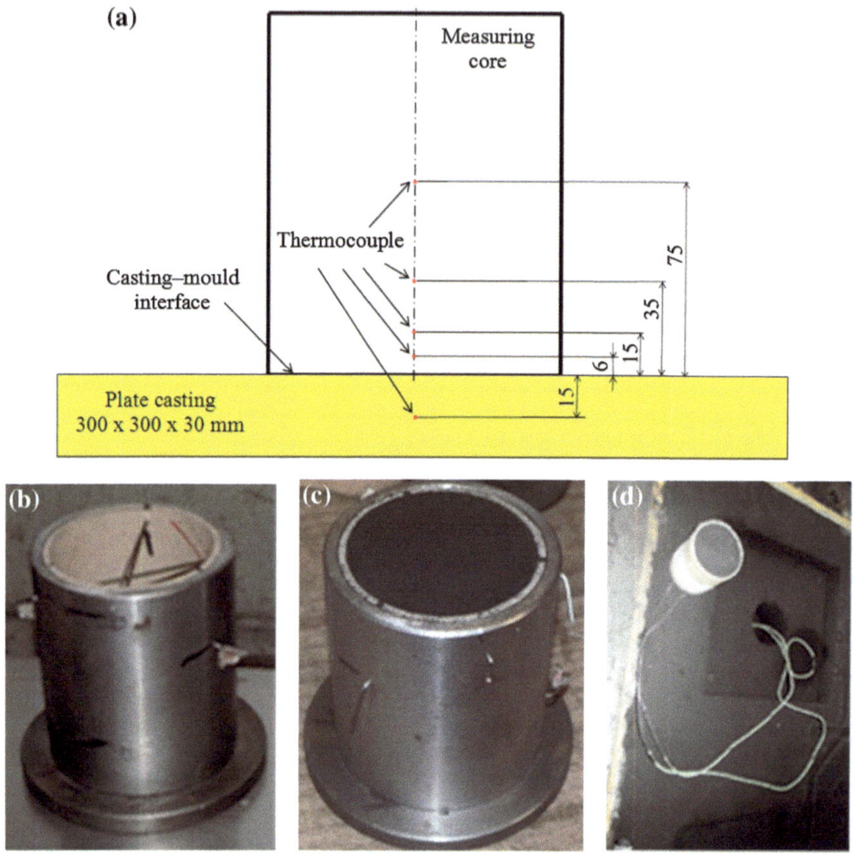

Fig. 6 Measuring core. **a** Scheme of instrumentation, **b**, **c** preparation and **d** assembly in the mould cavity

To determine the thermo-physical properties of mould sand containing the over-moisture zone an inverse problem solution was used. The originality of the study is an attempt to take into account the effects of the global thermal phenomena occurring in the quartz sand bonded by bentonite-water binder, using the apparent thermal coefficients. Numerical calculations were made by means of the casting simulation system (Procast) in which phenomena connected with the transport of humidity are not directly modeled. In order to include in the simulation the thermal effect of processes progressing deep into the mould, i.e. evaporation, vapour transfer and its condensation, the mould was divided into two zones:

- initial evaporation zone A (from mould-casting interface arbitrary defined layer of 4 mm for casting 30 mm thick)—characterized by substitute specific capacity ($C_{subst/modif//vapour} = c_{subst/modif//vapour} \cdot \rho_{subst}$) takes into account the latent heat of water vaporization),

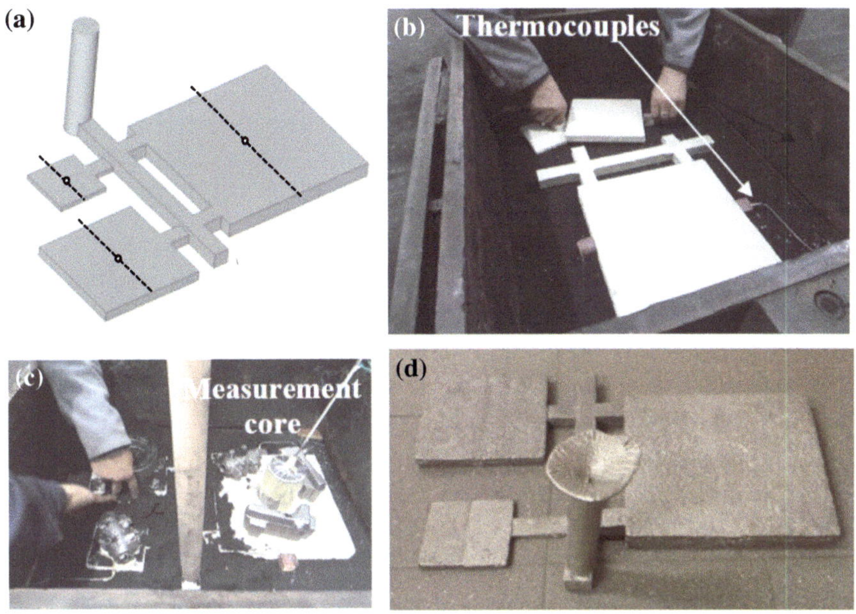

Fig. 7 Second experiment (Exp. 2). Casting arrangement in the mould. **a** CAD casting model (with thermocouples positions), **b**, **c** expanded polystyrene model and mould instrumentation, **d** raw casting knocking out and blasted

- hypothetic over-moisture zone B (outside of initial evaporation zone A > 4 mm from interface—characterized by the condensation latent heat (as direct internal heat source) and also substitute specific capacity (product of substitute specific heat and average density) takes into account the latent heat of water vaporization).

The adopted initial vaporization zone (endothermal character of vaporization heat) impacts the casting in the initial phase of its cooling process by increasing heat extraction from the casting. It results from the increased substitute thermal capacity of the mould. In the remaining part of the mould, the so-called hypothetic over-moisture zone there is a heat source of condensation. The heat of condensation (exothermic phenomenon) is generated in this zone as the function of the source when the temperature of the mould is below 100 °C. Over time, the influence of this zone on the speed of cooling the mould is increasingly lower. It was assumed that the sought humidity of moulding sand in the hypothetic over-moisture zone is constant (greater than initial humidity) and the heat of condensation is lower than the heat of vaporization by the value corresponding to the heat of water evaporation in sand with initial humidity (initial evaporation zone).

The simulation model simplifications (mould partition zones) are presented in Fig. 8.

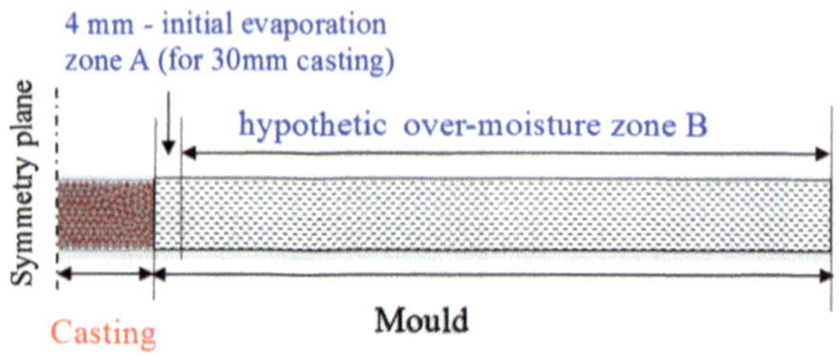

Fig. 8 Conventional partition of mould zones: zone A (humidity 3.5 %), zone B (constant humidity determined by inverse solution)

The simulation studies were conducted using the Procast system, according to the following scheme:

- virtual thermocouples were placed as in the experiment,
- simulation of filling, solidification and cooling to the final casting temperature 600 °C was carried out,
- validation was based on assumptions resulting from the scheme presented in the Fig. 9:

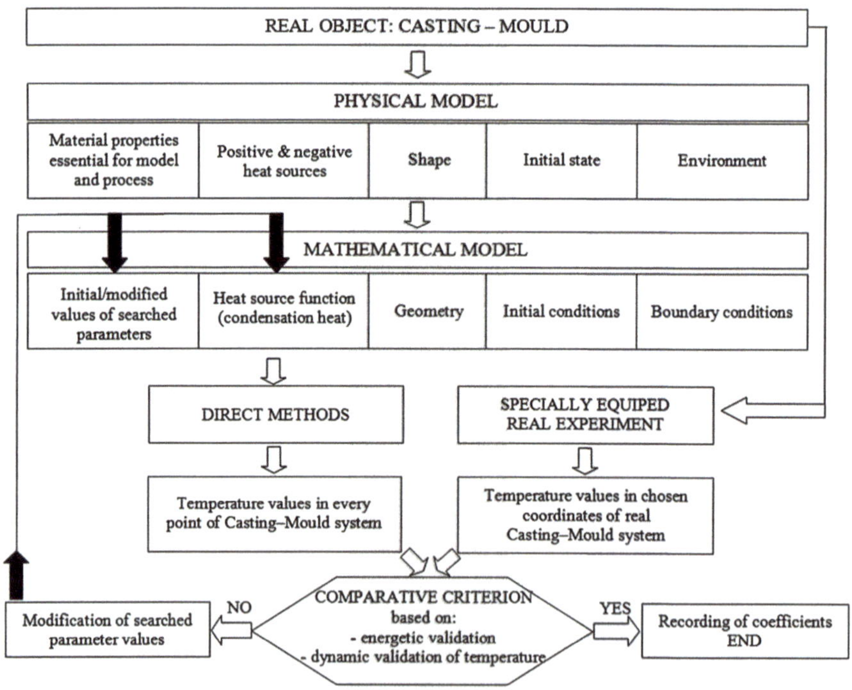

Fig. 9 Scheme of energetic/dynamic validation with the comparison criterion

1. energetic validation—on the basis of comparative analysis of solidification times determined from the cooling curves in geometric/thermal center of the casting,
2. dynamic validation of temperature—based on comparative analysis of the heating curves, recorded in the predefined mould sand points.

4 Results of Experimental Studies

Results of first experimental study (Exp. 1) in form of cooling curves of castings and their derivatives, along with curves of heating in the mould are presented in the Fig. 10. On the basis of the local minimum value of the first derivative, the plate solidification times were determined. Times of solidification were used to carry out the energetic validation. On the basis of the measurements of the real thermocouple locations in the mould sand, appropriate corrections were made after casting knocking out. It was taken into consideration in the inverse problem calculations.

Figure 11 shows cooling curves for the test casting and heating curves of the mould recorded in the second experiment (Exp. 2). Comparing the results of Exp. 1 and Exp. 2 for the plate 30 mm thick one can state that:

- in experiments Exp. 1 and Exp. 2 comparable solidification time of castings was obtained (solidification time about 800 s.)
- similar arrest lengths in temperature of about 100 °C corresponding to over-moisture zones were obtained,
- the mould of higher initial humidity heats up faster. In the case of a thermocouple positioned 35 mm from the interface casting-mould in experiment Exp. 1 the temperature of 100 °C was obtained after about 100 s, whereas in experiment Exp. 2 after about 220 s.

5 Results of Simulation Studies Including Inverse Solution

For the validation of the developed simplified model, the conditions of experiment Exp. 1 were adopted, in which the creation of the over-moisture zone for moulding sand of higher initial humidity (3.5 %) was tested.

Applying the rule resulting from the scheme presented in the Fig. 9, thermo-physical parameters of the moist mould sand in initial evaporation zone and in over-moisture zone were determined. In these parameters, phenomena of appearance and movement of the over-moisture zone in the mould sand are taken into account. In initial evaporation zone and over-moisture zone the specific heat

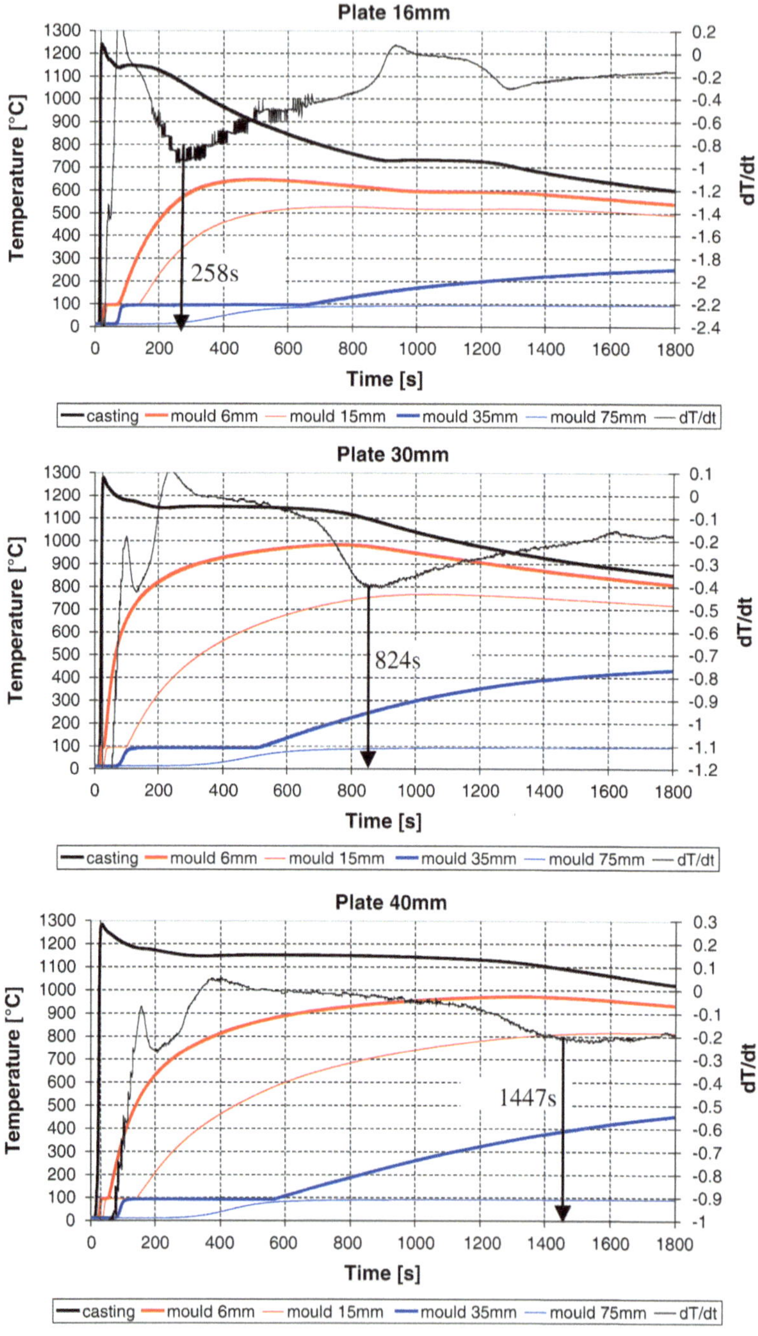

Fig. 10 Results of the experimental studies Exp. 1 (with estimation of real solidification times)

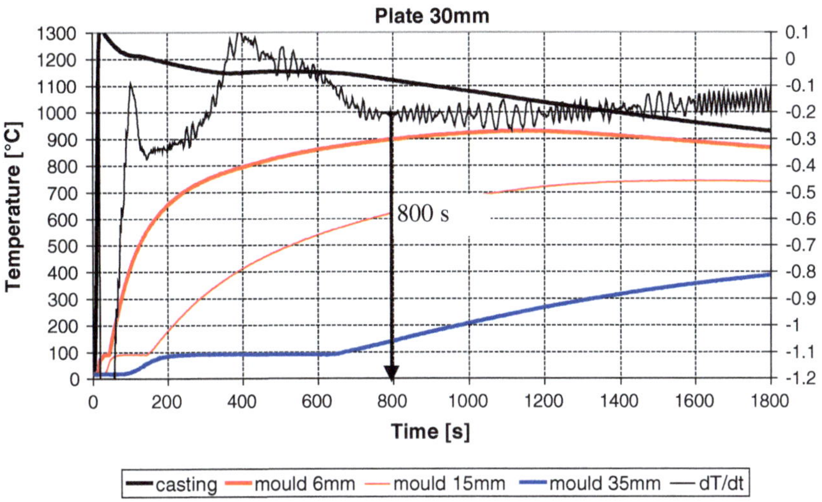

Fig. 11 Results of the experimental studies Exp. 2. Plate 30 mm

was modified, considering heat of evaporation of the water contained in the sand, assuming temperature range of 96–106 °C. The parameters determined during study of the green sand mould are:

- λ_{subst} = 0.8 W/(m.K)
- ρ_{subst} = 1540 kg/m^3
- $c_{subst/modif/vapour}$—presented in the graph of relation between the substitute specific heat and the temperature for the tested sand mould (Fig. 12).
- the condensation latent heat in over-moisture zone was determined as the heat source function (according to Procast algorithm) with following parameters: condensation energy 474 kJ/kg, start temperature 50 °C and time of reaction—10 s.

As a result of the validation studies, carried out according to the scheme shown in the Fig. 9, satisfying compatibility between the best curves from the simulation and from the experiment was achieved by simplified inverse solution (trial and error method). The results of the final simulation are shown in the Fig. 13. The good agreement for final case between times of solidification in the experiment (Fig. 10) and in the simulation was obtained. Also in the Fig. 13 good agreement of temperature arrest about 100 °C is observed (compare Fig. 10).

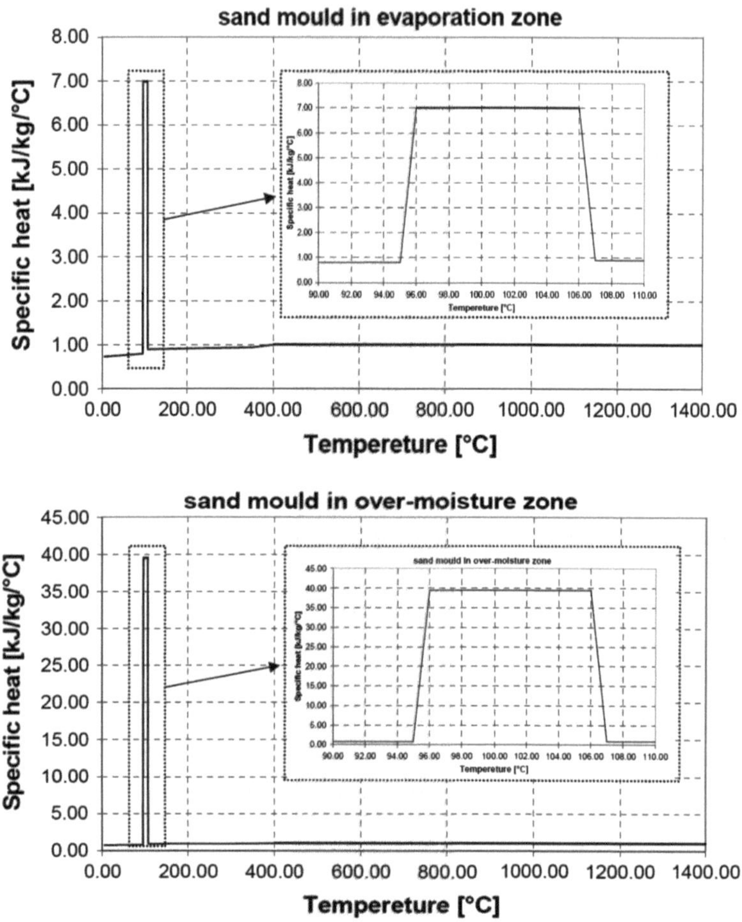

Fig. 12 Substitute specific heat of the green sand mould determined by inverse solution (assumption: for average ρ_{subst} = 1540 kg/m^3)

6 Application of Green Sand Mould Parameters Determined in This Study

The thermo-physical parameters of the green mould sand determined on the basis of the described validation studies were used in a simulation of castability trial [37] for cast iron EN-GJL-250 (look Table 1—casting process parameters). The castability of cast iron depends on fluidity of metal poured into standard mould cavity made of tested sand. In foundry practice this is the ability of alloy to flow through the gating system and filling the casting walls. The castability value as complex property combining both alloy and mould characteristics, is quantitatively defined by

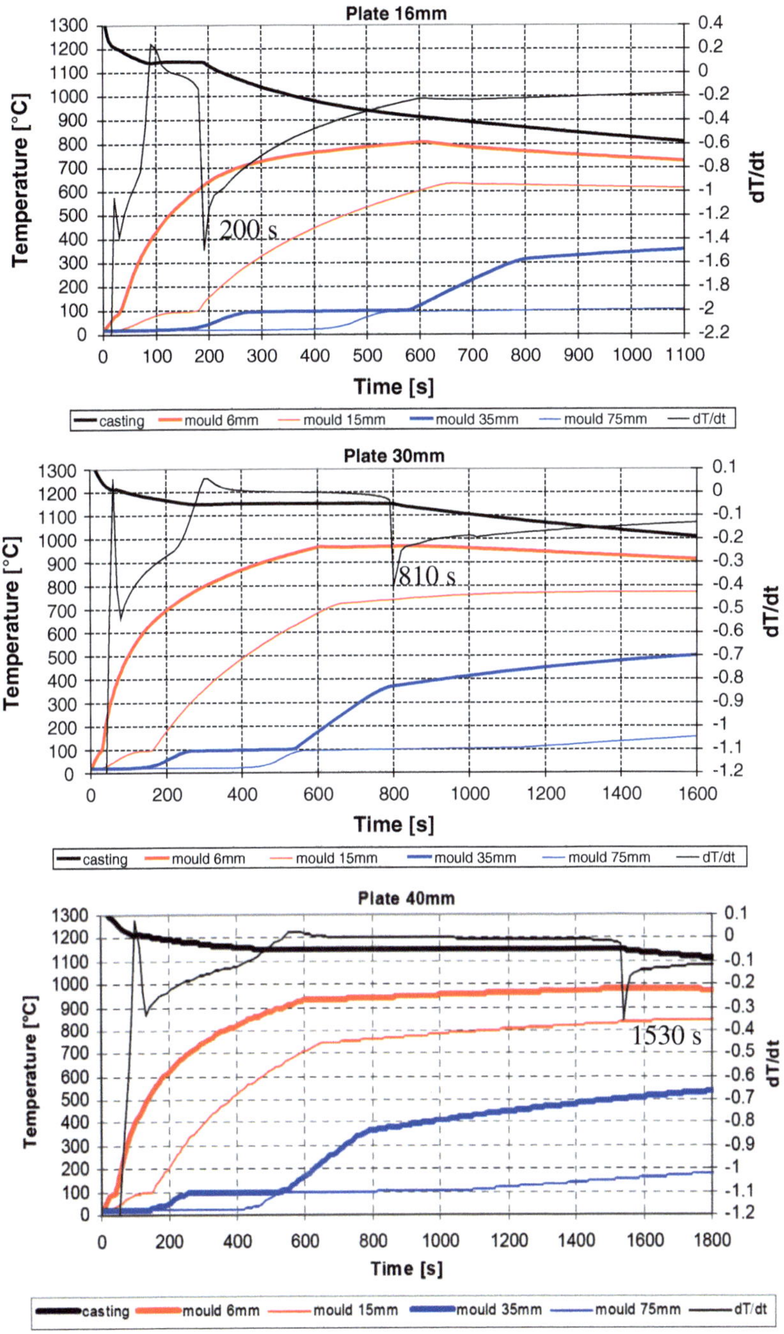

Fig. 13 Results of the simulation studies (with estimation of virtual solidification times based on the characteristic minimum values of cooling curves derivative—d*T*/d*t*)

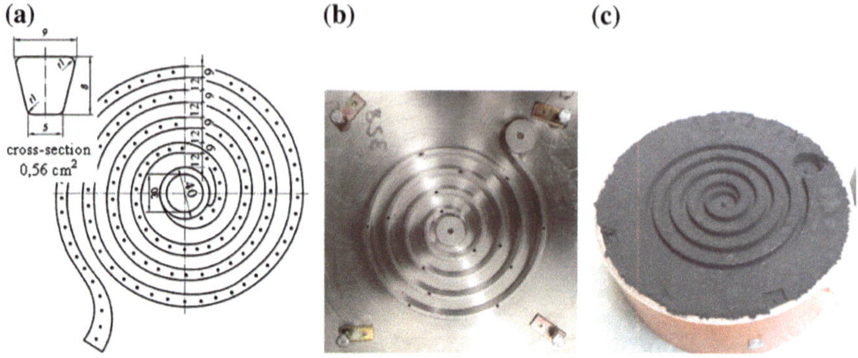

Fig. 14 Preparation of castability trials mould, **a** castability trial (*spiral shape*), **b** metallic pattern plate, **c** view of the mould (castability trial spiral test)

maximum length of the obtained casting (often on spiral shape). The filling of spiral canal is done without a refresh of stream front which is stopped by the local crystallization on the front region.

Three verifying experiments of casting castability spirals for various initial conditions (various temperatures of pouring) and for various types of cast iron were carried out. The course of the preparation of the casting experiment is shown in Fig. 14 and the results of fluidity tests of casting spirals in Fig. 15.

Next, a series of simulations for the castability trial casting process was performed. The results of the simulation were compared with the experiment test T1 (Fig. 15a). The influence of substitute thermo-physical parameters of moulding

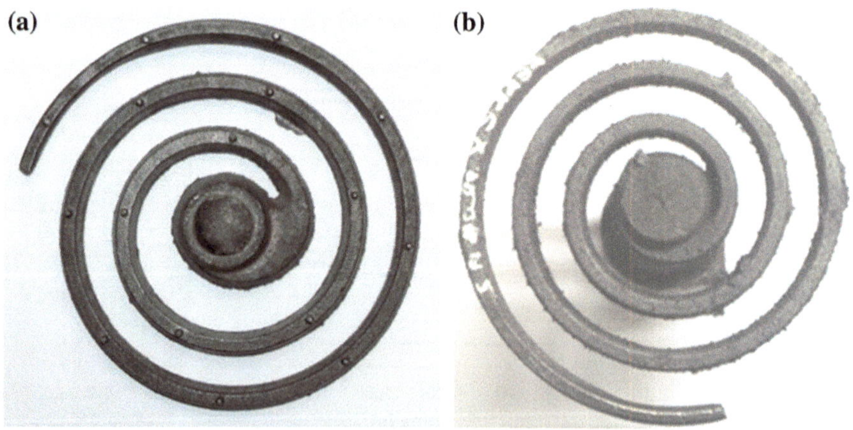

Fig. 15 Castability trial (*spiral shape*)—examples of test castings: **a** result of test T1, pouring temperature 1345 °C, **b** result of test T2, pouring temperature 1367 °C

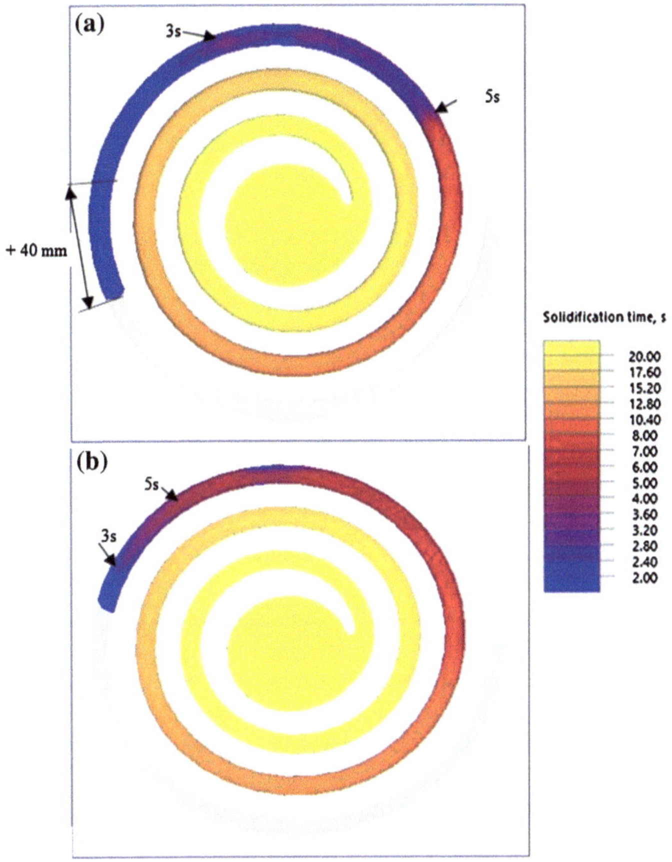

Fig. 16 Castability trial—the comparison of filling simulation results (with the test casting T1 Fig. 15a): **a** simulation using dry mould sand parameters from the NovaFlow & Solid database, **b** simulation using mould sand parameters (specific heat and heat source function) determined by the validation studies

sand in the over-moisture zone on the length of a fluidity spiral forecast by simulation was observed. The results of two simulations of pouring process were compared; in the first one, data for the green sand was taken from the database of the NovaFlow & Solid system (dry green sand), in the second one—data determined in the described validation studies (described in Chap. 5) was used. The results of simulation are presented in the Fig. 16 which show an evident influence of presence of the over-moisture zone in the mould sand on filling of the castability trial result.

7 Conclusions

The main conclusions that we can draw are as follows:

- In this study the effective way to take into account in the foundry simulation system the presence of over-moisture zone and its impact on the cooling rate of the casting was found.
- The thermo-physical substitute parameters of the green sand mould in initial evaporation zone and in over-moisture zone were determined.
- The initial humidity of moulding sand impacts the dynamics of the movement of the over-moisture zone and heating the mould. Comparing the results of the experiment involving casting 1 and 2 for the plate 30 mm thick, one may state that in moulding sand with higher initial humidity (3.5 %) the over-moisture zone formed and moves faster that in sand with humidity level of 2.5 %. The impact of initial humidity on the length of the effect of the over-moisture zone was not observed (length of temperature arrests about 100 °C).
- Average mould humidity in over-moisture zone, determined on the basis of the inverse problem solution, is about 21 % (similar to the value from bibliography).
- Using the substitute thermo-physical parameters of green sand mould, elaborated by inverse problem solution, the satisfied agreement of calculated solidification times for plates castings was obtained (energetic validation condition).
- On the mould heating simulation curves, the effect of presence of the over-moisture zone in form of long range temperature arrest (around 100 °C) is observed. The length of simulated arrest is comparable with real experiment curves (proved by dynamic temperature validation).
- The proposed evaporating zone with initial water content (few millimetres thick depending of casting wall thickness) in simulated process increases the heat capacity of the mould, which results in increased extraction of heat from the casting during initial phase of the casting process.
- The elaborated in described study data allowed to conduct a virtual experiment of castability trial near to real experiment. In such a way, a necessity of taking into consideration the parameters allowing recreation of the over-moisture zone in the model was confirmed, when the cast iron casting is manufactured in the green sand mould. It is of great significance especially during simulation of filling of the canals of the gating system and/or thin walls of the casting.

References

1. Bogosłowski, W.: Hygrothermal Problems of Buildings (in Polish). Arkady, Warszawa (1989)
2. Kasperkiewicz, J.: Moisture Transport and Shrinkage Deformation in Concrete (in Polish). PWN, Warszawa (1972)
3. Lesniewska, M., Pogorzelski, A.: Study of rising damp in building materials (in Polish). Arch. Inż. Ląd. **2**(22), 333–343 (1976)

4. Guimarães, A., de Freitas, V.P., Delgado, J.M.P.Q.: Treatment of Rising Damp in Historical Buildings. Heat and Mass Transfer in Porous Media. Springer, Berlin (2011)
5. Gawin, D.: Modelling of coupled hygro-thermal phenomena in building materials and building components. Zeszyty Naukowe Nr 853, Wydawnictwo Politechniki Łódzkiej, Łódź (2000)
6. Garbalińska, H.: Isothermal Coefficients of Moisture Transport in Porous Building Materials (in Polish). Szczecin (2002)
7. Ickiewicz, I., Sarosiek, W., Ickiewicz, J.: Selected Topic in Building Physics (in Polish). Białystok (2000)
8. Kubik, J.: Moisture Transport in Building Materials (in Polish). Politechnika Opolska (2000)
9. 48th Census of World Casting Production, Modern Casting Dec/2014
10. Raport CAEF—The European Foundry Industry (2013)
11. Sobczak, J., Balcer, E., Kryczek, A.: Situation of the foundry engineering in Poland and in the world—the current state and predictions. Przegląd Odlewnictwa, nr 1–2 (2012)
12. https://www.esi-group.com/software-services/virtual-manufacturing/casting/procast-quikcast
13. http://novacast.se/products/novaflowsolid/
14. Szmigielski, T.: Moisture of mould in condensation zone. Arch. Foundry **3**(10), 249–254 (2003)
15. Szmigielski, T.: Phenomena in over-moisture zone in wet sand mould heated on one side. In: Proceedings of X Foundry Conference TECHNICAL (in Polish), 2007
16. Szmigielski, T.: New version of a stand for testing resistivity of damp moulding sands. Arch. Foundry **6**(22), 526–531 (2006)
17. Szmigielski, T.: Resistant characteristics of damp moulding sand. Arch. Foundry **4**(14), 508–513 (2004)
18. Smolinski, A., Pucka, G., Rojek J.: Technological aspects of the over—moisture zone in clay moulding sands. Arch. Foundry **4**(14), 450–455 (2004)
19. Chowdiach M.P.: Giesserei 19, 582–590 (1971)
20. Marek, C.T.: Modern Casting, nr 2 (1968)
21. Rzeczkowski, M.: Analysis of Water Circulation Phenomena in the Green Sand Moulds Influenced by Temperature (in Polish). Scientific Letters of WSInz (1977)
22. Ignaszak, Z., Graczyk, L., Popielarski, P.: Experimental—simulating method of over moist zone identification in the mould. Arch. Foundry **6**(22), 216–223 (2006)
23. Chu, Shao-Shu, Fang, Te-Hua, Chang, Win-Jin: Modelling of coupled heat and moisture transfer in porous construction materials. Math. Comput. Model. **50**, 1195–1204 (2009)
24. Guimarães, A.S., Delgado, J.M.P.Q., de Freitas, V.P.: Numerical simulation of rising damp phenomenon. Defect Diffus. Forum **326–328**, p48–p53 (2012)
25. Delgado, J.M.P.Q., Vázquez da Silva, M.: Performance and modelling of water vapour adsorption in piles of granules using a cylindrical pore model. Defect Diffus. Forum **312–315**, 1155–1160 (2011)
26. Garbalińska, H.: Isothermal Coefficients of Moisture Transport in Porous Building Materials (in Polish). Szczecin (2002)
27. Werner, H., Gertis, K.: Energetische Kopplung von Feuchte- und Wärmeübertragung an Außenflächen, In: Hygrische Transportphänomene in Baustoffen, Deutscher Ausschuß für Stahlbeton, Heft 258, Berlin (1976)
28. Huang, C., Ye, H., Fan, J., Sun, W.: Numerical study of heat and moisture transfer in textile materials by a finite volume method. Commun. Comput. Phys. **4**(4), 929–948 (2008)
29. Fan, Jintu, Wen, Xinghuo: Modelling heat and moisture transfer through fibrous insulation with chase change and mobile condensates. Int. J. Heat Mass Transf. **45**, 4045–4055 (2002)
30. Fan, J., Luo, Z., Li, Y.: Heat and moisture transfer with sorption and condensation in porous clothing assemblies and numerical simulation. Int. J. Heat Mass Transf. **43**, 2989–3000 (2000)
31. Fan, J., Wen, X.: Modelling heat and moisture transfer through fibrous insulation with phase change and mobile condensates. Int. J. Heat Mass Transf. **45**, 4045–4055 (2002)
32. Farnworth, B.: Mechanics of heat flow through clothing insulation. Tex. Res. J. **53**(12), 717–725 (1983)
33. Handbook of Thermophysical Properties (ed.) Macmillan Comp., New York (1961)

34. Ignaszak, Z.: Thermophysical Properties of Mould Materials in Aspect of Solidification Control of Simulation Process (in Polish). Poznan University of Technology, Rozprawy nr 211, Poznan (1989)
35. Popielarski, P., Ignaszak, Z.: Modelling capabilities of phenomena in over-moisture zone existing in porous medium using the simplified simulation systems applied in foundry. In: Proceedings of the 10th International Sympodium on the Science and Processing of Cast Iron (SPCI10), Argentina, 2014
36. http://www.magmasoft.com/en/
37. BN-79/4051-17—Polish Industy Standard: Technological Castability Trial for Alloys

Optimization of Pulsed Vacuum Osmotic Dehydration of Sliced Tomato

J.L.G. Corrêa, A. Dantas Viana, K. Soares de Mendonça and A. Justus

Abstract The osmotic dehydration (OD) is a treatment that reduces partially the moisture content and the water activity of a food. The use of reduced pressure in the first minutes of an OD is called pulsed vacuum osmotic dehydration (PVOD). In PVOD, the expulsion of occlude gases and the entrance of the solution in the food matrix are induced, with consequently mass transfer improvement. In the present work, the influences of independent variables (temperature (T), pressure of vacuum pulse (PV), sodium chloride concentration [NaCl] and sucrose concentration [suc]) on water loss (WL), solid gain (SG), weight reduction (WR), water activity (a_w) and moisture content (U) of tomato slices were studied. The optimum condition obtained had its kinetics tested with three models from the literature. The optimized condition (T 40 °C, PV 56.25 mbar, [NaCl] 7.5 % and [suc] 32.5 %) resulted in the maximum values of WL, and WR (42.2, and 36.1 %, respectively) and minimum SG, a_w and U (4.0, 0.948 and 76.5 kg water/100 kg sample, respectively). With respect to the kinetics, the best agreement was obtained with the model that considers variable diffusivity.

J.L.G. Corrêa (✉) · K.S. de Mendonça · A. Justus
Universidade Federal de Lavras, Departamento de Ciência dos Alimentos, Lavras, Brazil
e-mail: jefferson@dca.ufla.br

K.S. de Mendonça
e-mail: keamendonca@msn.com

A. Justus
e-mail: arianajustus@yahoo.com.br

A. Dantas Viana
Universidade Federal da Paraíba, Departamento de Gestão e Tecnologia Agroindustrial, João Pessoa, Brazil
e-mail: arianneviana@hotmail.com

J.M.P.Q. Delgado and A.G. Barbosa de Lima (eds.),
Drying and Energy Technologies, Advanced Structured Materials 63,
DOI 10.1007/978-3-319-19767-8_11

List of Symbols

Symbol

OD	Osmotic dehydration
PVOD	Pulsed vacuum osmotic dehydration
T	Temperature (°C)
PV	Pressure of vacuum pulse (mbar)
[NaCl]	Sodium chloride concentration (%)
[suc]	Sucrose concentration (%)
WL	Water loss (kg water/100 kg sample)
WR	Weight reduction (kg/100 kg sample)
SG	Solid gain (kg solid/100 kg sample)
a_w	Water activity
U	Moisture content (kg water/100 kg sample)
w	Weight (kg)
x^w	Water content (%)
x^{ST}	Solid content (%)
X_1	Independent variable (T)
X_2	Independent variable (PV)
X_3	Independent variable ([NaCl])
X_4	Independent variable ([suc])
z	Specific directional coordinate (m)
t	Time (s)
m	Amount of water or solids (kg)
D_{eff}	Effective diffusion coefficient ($m^2\ s^{-1}$)
L	Half thickness of sample (m)
i	Number of series terms
W	Dimensionless water or solid content
z^{ss}	Mass fraction of soluble solids in the sample (kg/100 kg sample)
y^{ss}	Mass fraction of soluble solids in the osmotic solution (kg/100 kg osmotic solution)
Y	Drive force
k_1	Adjusted Peleg's parameter
k_2	Adjusted Peleg's parameter
X	Water loss or solid gain (kg/100 kg sample)
$t_{1/2}$	Half-life of dehydration/impregnation rate (min)
SE	Standard error

Subscripts

0	Initial condition
f	Final condition
eq	Equilibrium condition
w	Water
s	Solid

HDM Hydrodynamic mechanism
∞ At infinite time

1 Introduction

The tomato is produced and consumed worldwide, in fresh or processed way. Its world production was 1.61×10^8 t in 2012 [22]. Dehydrated products are interesting by their larger shelf life, reduced volume and weight, value added and availability [9, 10, 21, 33, 43]. In this regard, dehydrated tomato is used as ingredient of soaps, pizzas and bakery goods. Besides, it is a great source of lycopene with nutritional acceptance [28].

Osmotic dehydration (OD) corresponds to the immersion of the food in a hypertonic solution with two counter and simultaneous diffusional fluxes: water from the food to the solution and soluble solids, from the solution to the food [41]. A third, but less expressive flux correspond to native soluble solids from the food to the solution. Even though the product from an OD is partially dehydrated, it still presents large water activity [15, 16, 54] and should be submitted to other food conservation process, as drying [13, 14, 21, 49, 55]. The partial moisture reduction could represent an overall energy saving because it improves the further drying rate [60]. Moreover, higher drying rate means reduced drying time, i.e., a shorter period in contact with heated air is required for obtaining a product with the same moisture content. This is an important aspect for food that contains thermal sensible nutrients [60].

The OD is a diffusive process. The water and solids diffusivities are function of process variables as solution concentration and viscosity, temperature and pressure and solution-food contact [59]. Material characteristics as porosity and texture are also relevant [38, 57]. Standard OD is usually performed at atmospheric pressure and room temperature [55]. However, the use of reduced pressure in the first minutes could be addressed, in a process called pulsed vacuum osmotic dehydration or PVOD [15, 16, 20, 21, 56]. The vacuum promotes the removal of internal gases located in the intercellular space and the entrance of the osmotic solution in the food matrix. This exchange, induced by the low pressure, is called hydrodynamic mechanism (HDM) and brings a non-diffusive aspect to the process [12].

The osmotic process is time dependent. Its experimental kinetic data is generally fitted by the diffusive model of Crank [18] or a modification of that model [11, 16, 31, 52]. It is also found empiric relations [5, 11, 19], probabilistic models [17] and numerical methods [40, 50, 51]. The consideration of constant diffusivity in the model of Crank [18] could result in a lack of fit. Some models consider variable diffusivity [8, 11] and could transpose the lack of precision of Crank model.

OD is largely employed as a pretreatment for the dried tomatoes [3, 13, 48, 52]. This pretreatment could preserve the typical aroma, improving the quality of the final product [26, 28]. Traditionally, OD is performed at atmospheric pressure [13, 44] with sucrose or sodium chloride solution or in a ternary solution of sucrose-sodium chloride [2, 3, 30, 31, 52, 53]. The ternary solution of sucrose-sodium chloride was recommended by its high efficiency in removing water and lycopene reserving [25–28]. The dried cherry tomato pretreated by PVOD in sucrose solution presented, in comparison with OD, decreased color change and hardness and increased vitamin C retention [3]. The present work aimed to obtain the optimized operational condition of PVOD of tomatoes (*Lycopersicon esculentum*, cultivar Carmen) in a ternary solution of sucrose-sodium chloride with respect to mass transfer parameters. The fitness of PVOD kinetics was also tested by three different models from the literature.

2 Materials and Methods

The methodology and the material used are described below.

2.1 Material

Tomatoes (*Lycopersicon esculentum*) cultivar Carmen were purchased in the local market (Lavras, MG, Brazil). The fruits were chosen visually by size, weight, color intensity and uniform firmness, to be considered homogeneous.

The raw material was characterized according to its centesimal composition in relation to moisture content, ashes, lipids, proteins, fibers [4] and carbohydrates, as follows: moisture content 94.98 ± 0.20 kg/100 kg, lipids 0.12 ± 0.01 kg/100 kg; proteins 0.04 ± 0.00 kg/100 kg; fibers 1.24 ± 0.13 kg/100 kg; ashes 0.46 ± 0.10 kg/100 kg and carbohydrates 2.25 ± 0.01 kg/100 kg. The water activity (Aqualab, model CX-2T Decagon Devices Inc., Pullman, WA, USA, 25 °C) was 0.996 ± 0.001.

2.2 Preparation of the Samples and Osmotic Solution

Tomatoes present waxy epidermis that could makes difficult mass transfer between the fruit and the immersion media [1]. The epidermis presents high lycopene content. Due to this fact, the epidermis was preserved [58]. The mass transfer in the epidermis was provided by immersing the whole fruit in a 5 g/100 g sodium hydroxide solution for 15 min at ambient temperature (25 °C). This treatment caused fissures in the tomatoes epidermis [32, 44]. After this, the tomatoes were washed in tap water, cut into halves and had the seeds removed. The samples used in the experiments were obtained with the aid of a stainless steel shaper. They were flat slices with the dimensions (4.0 ± 0.2) cm length, (2.0 ± 0.2) cm width and the thickness of the fruit pulp, (0.8 ± 0.2) cm.

The hypertonic solution was composed of distilled water, commercial sodium chloride and commercial sucrose. The composition and concentration of the solution for each run was in accordance with the levels presented on Table 1.

Table 1 Actual and coded values of independent variables used for experimental design

Run	Temperature (°C)		Vacuum pressure (mbar)		Sodium chloride concentration (%)		Sucrose concentration (%)	
	Actual, T	Coded, X_1	Actual, PV	Coded, X_2	Actual, [NaCl]	Coded, X_3	Actual, [Suc]	Coded, X_4
1	30	−1	37.5	−1	2.5	−1	17.5	−1
2	40	+1	37.5	−1	2.5	−1	17.5	−1
3	30	−1	112.5	+1	2.5	−1	17.5	−1
4	40	+1	112.5	+1	2.5	−1	17.5	−1
5	30	−1	37.5	−1	7.5	+1	17.5	−1
6	40	+1	37.5	−1	7.5	+1	17.5	−1
7	30	−1	112.5	+1	7.5	+1	17.5	−1
8	40	+1	112.5	+1	7.5	+1	17.5	−1
9	30	−1	37.5	−1	2.5	−1	32.5	+1
10	40	+1	37.5	−1	2.5	−1	32.5	+1
11	30	−1	112.5	+1	2.5	−1	32.5	+1
12	40	+1	112.5	+1	2.5	−1	32.5	+1
13	30	−1	37.5	−1	7.5	+1	32.5	+1
14	40	+1	37.5	−1	7.5	+1	32.5	+1
15	30	−1	112.5	+1	7.5	+1	32.5	+1
16	40	+1	112.5	+1	7.5	+1	32.5	+1
17	25	−2	75	0	5	0	25	0
18	45	+2	75	0	5	0	25	0
19	35	0	25	−2	5	0	25	0
20	35	0	45	+2	5	0	25	0
21	35	0	75	0	0	−2	25	0
22	35	0	75	0	10	+2	25	0
23	35	0	75	0	5	0	10	−2
24	35	0	75	0	5	0	40	+2
25	35	0	75	0	5	0	25	0
26	35	0	75	0	5	0	25	0
27	35	0	75	0	5	0	25	0
28	35	0	75	0	5	0	25	0

2.3 Experimental System

The process of osmotic dehydration was performed in an osmotic dehydrator designed to work at atmospheric or reduced pressure (see Fig. 1). The device [15, 56] is a stainless steel chamber with control of temperature, agitation and inner pressure. The internal volume of the dehydration is 50 L, minimum operational volume of 10 L. It is covered by a jacket with water as thermal fluid. A vacuum pump coupled to the bath allows the obtainment of reduced pressure. The homogeneous composition and temperature is acquired with the aid of an impeller in the bottom of the chamber.

2.4 Osmotic Dehydration

The hypertonic solution was introduced in the dehydration at the set concentration and temperature (see Table 1). The samples, placed in stainless steel perforated

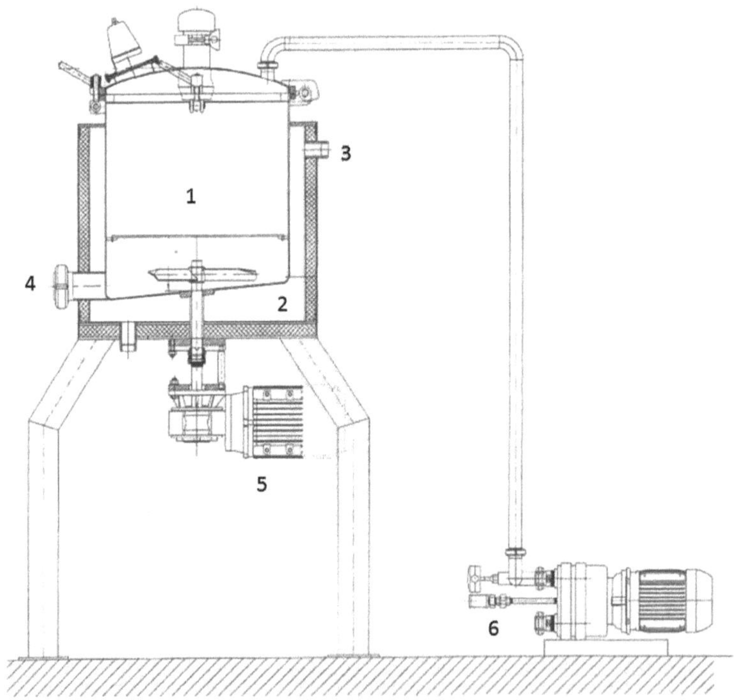

1 Internal volume; 2 External volume; 3 Exit to thermostatic bath; 4 Osmotic solution output; 5 Agitation impeller system; 6 Vacuum pump

Fig. 1 Experimental setup. *1* Internal volume; *2* jacket volume; *3* exit to thermostatic bath; *4* osmotic solution output; *5* agitation impeller system; *6* vacuum pump boxes, were, then, immersed in the solution in a fruit:solution ratio of 1:59 (w:w) to avoid dilution of the solution during the osmotic process with consequent attainment of the driving force during the process [23]. The boxes allowed, for each sample, identification and immersion in a single layer. For experiments at reduced pressure, called pulsed vacuum osmotic dehydration (PVOD), there was application of vacuum in the first 10 min of the process. The vacuum pressure is according to Table 1 and the time, optimized by previous experiments. After this period, the process continued at atmospheric pressure.

The samples were taken off the solution and from the boxes at predetermined times. They were, then, immersed in a bath of ice and distilled water for until 10 s to stop the dehydration and remove the excess of solution from their surface. After this, the surface of the samples was carefully dried with absorbent paper towel. The samples had their mass and moisture content measured. The water loss (WL), solid gain (SG) and weight reduction (WR) were calculated in accordance with Eqs. (1)–(3), respectively and the kinetics of moisture content (U), WL, SG and WR, obtained.

$$WL\,(\%) = \frac{x_0^w M_0^o - x_f^w M_f^o}{M_0^o} 100 \tag{1}$$

$$SG\,(\%) = \frac{x_f^{ST} M_f^o - x_0^{ST} M_0^o}{M_0^o} 100 \tag{2}$$

$$WR(\%) = \frac{M_0^o - M_f^o}{M_0^o} 100 \tag{3}$$

where M_0^o = initial weight of the sample (kg), M_f^o = final weight of sample (kg), x_0^w = initial water content (%), x_f^w = final water content (%), x_0^{ST} = initial solid content (%), and x_f^{ST} = final solid content (%).

2.5 Experimental Design and Optimization

The strategy for design the optimum condition of the PVOD was based in a central composite rotatable design (CCRD) [42]. The independent variables were temperature (T, X_1), pressure of vacuum pulse (PV, X_2), sodium chloride concentration ([NaCl], X_3) and sucrose concentration ([Suc], X_4). The dependent variables were water loss (WL), solid gain (SG), weight reduction (WR), water activity (a_w) and moisture content (U). Table 1 presents the real and coded variables for each run.

The best condition was determined using response surface methodology, which was performed by analyzing the response surface of WL, SG, WR, a_w and U, to maximize WL and WR and minimize SG, a_w and U.

2.6 Kinetic Models

The experimental kinetics data of the WL and SG at the best condition were fit by three mathematical models with estimation of the water and the solids diffusion coefficients. The tested models were the unidirectional diffusion model, based on Fick's 2nd law [18], the hydrodynamics model of Fito and Chiralt [24] and the model of Barbosa Júnior et al. [8].

2.6.1 Diffusion Model of Crank

The diffusion model of Crank [18] is based in the second law of Fick. It considers the diffusion of a compound in a specific directional coordinate, z, with time, t as:

$$\frac{\partial m(t)}{\partial t} = \frac{\partial}{\partial z}\left(D_{eff}\frac{\partial m(t)}{\partial z}\right) \tag{4}$$

where m(t) is the amount of the compound, water or solids, in the case, at the instant t and D_{eff}, the effective diffusion coefficient. The solid sample was regarded to be a 2-L-thick plate. The amount is considered uniform at t = 0 $m_{(z,0)} = m_0$; and the symmetry of concentration, $\left.\frac{\partial m(t)}{\partial t}\right|_{z=0} = 0$ is considered. Finally, the equilibrium is considered at the material surface, $m_{(L,t)} = m_{eq}$.

Equation 4, solved with the considerations [18] becomes:

$$W = \left(\frac{8}{\pi^2}\sum_{i=1}^{\infty}\frac{1}{(2i+1)^2}\exp\left(-(2i+1)^2\pi^2 D_{eff}\frac{t}{4L^2}\right)\right) \tag{5}$$

where D_{eff} is the effective diffusion coefficient of the water or soluble solids, i is the number of series terms and W is the dimensionless water or solid content, given by Eq. (6):

$$W = \frac{m(t) - m_{eq}}{m_0 - m_{eq}} \tag{6}$$

The term m_{eq} was experimentally obtained for samples without mass variation after more than 48 h of OD.

2.6.2 Hydrodynamics Model

The hydrodynamics model of Fito and Chiralt [24] corresponds to Eqs. (7)–(13). This mathematical model regards an equilibrium approach (Eq. 7):

$$z_{eq}^{SS} = y_{eq}^{SS} \tag{7}$$

where z_{eq}^{SS} is the mass fraction of the soluble solids in the food and y_{eq}^{SS} is the mass fraction of the soluble solids in the osmotic solution, both at the equilibrium state. As a consequence, the effective diffusion coefficient (or pseudo diffusivity) is the same for both water and solids:

$$D_{eff_{w\,or\,s}} = D_{eff_w} = D_{eff_s} \tag{8}$$

The variations in the composition are functions of the reduced drive force, Y, given by:

$$Y = Y_t^w = Y_t^s = \frac{z_t^w - z_{eq}^w}{z_0^w - z_{eq}^w} \tag{9}$$

The change on the food liquid phase composition connected to the hydrodynamic mechanism (HDM) happens at the very beginning of the process (t = 0 to t = t_{HDM}), where this mechanism is prevalent and is dependent on the pressure gradients:

$$1 - Y_t^w\Big|_{t=0}^{t=t_{HDM}} \cong k \tag{10}$$

After this period, the phenomena are modelled with Fick's equation for semi-infinite slab and short time [18], with the approach suggested by Fito and Chiralt [24]:

$$1 - Y_t^w\Big|_{t=t_{HDM}}^{t=t} = 2\left(\frac{D_{eff}t}{L^2}\right)\left(\pi^{-0.5} + 2\sum_{i-1}^{\infty} \mathrm{ierfc}\frac{iL}{\sqrt{D_{eff}t}}\right) \tag{11}$$

Equation 11 can be simplified and approximated by Eq. (12):

$$1 - Y_t^w\Big|_{t=t_{HDM}}^{t=t} = 2\left(\frac{D_{eff}t}{\pi L^2}\right)^{0.5} \tag{12}$$

These two effects were conjugated to consider the effect of the hydrodynamics and the pseudo-Fickian mechanisms:

$$1 - Y_t^w\Big|_{t=0}^{t=t} = k + 2\left(\frac{D_{eff}t}{\pi L^2}\right)^{0.5} \tag{13}$$

The D_{eff} and k parameters were obtained for each experiment from a linear fitting of the experimental $1 - Y_t^w\big|_{PD,t>0}$ versus $t^{0.5}$.

2.6.3 Modified Diffusion Model—Model of Barbosa Júnior et al.

With the goal of considering D_{eff} variable with time, Barbosa Júnior et al. [8] proposed a modification in the model of Crank. The Eq. (5) is rewritten as Eq. (14) (see [6, 18, 37]):

$$W = \left(\frac{8}{\pi^2}\sum_{i=1}^{\infty}\frac{1}{(2i+1)^2}\exp\left(-(2i+1)^2\pi^2 D^i_{eff}\frac{t}{4L^2}\right)\right) \tag{14}$$

where D_{eff}^{i} is the effective diffusion coefficient [$m^2\ s^{-1}$] considered as variable with time.

The equilibrium condition is obtained by the two-parameter, nonexponential, empirical equation proposed by Peleg [39]:

$$m^i_t = m^i_0 \pm \frac{t}{k^i_1 + k^i_2 t} \tag{15}$$

where m^i is the amount [kg] of solids (i = s) or water (i = w) at time t (m^i_t) or at the process beginning (m^i_0); $k_1{}^i$ and $k_2{}^i$ are the adjusted Peleg's parameters. In Eq. (15), '±' becomes '+' or '−' if the solid gain or the water loss are, respectively, evaluated. If the moisture content in wet basis, U, is taken into account:

$$m^w_t = m_t \times U_t \tag{16a}$$

$$m^s_t = m_t \times (1 - U_t) \tag{16b}$$

Equations 16a and 16b can be substituted into water loss (WL) and solid gain (SG) equations [16, 47, 56], allowing the Peleg's model to be rewritten to describe WL and SG as X (see Eq. 17):

$$X(t) = \frac{t}{k^i_1 + k^i_2 t} \tag{17}$$

For t → ∞, Eq. 15 is presented as Eq. (18) and k^i_2 is obtained:

$$X_{eq} = \lim_{t\to\infty}\left(\frac{t}{k^i_1 + k^i_2 t}\right) = \frac{1}{k^i_2} \tag{18}$$

For t = 0, Eq. (15) is presented as Eq. (19) and k^i_1 is obtained:

$$\left.\frac{dX(t)}{dt}\right|_{t=0} = \frac{1}{k^i_1} \tag{19}$$

Therefore, this indicates that a relationship exists between dehydration/impregnation rates and the driving force of the process. The *n*-reduction time for the dehydration/impregnation rate ($t^i_{1/2}$) is defined as follows:

$$\left.\frac{dX(t)}{dt}\right|_{t=t^i_{1/n}} = \frac{1}{n}\left.\frac{dX(t)}{dt}\right|_{t=0} \tag{20}$$

The process begins at n = 1, and n = 2 is the 'half-life' of rate ($t^i_{1/2}$), that is, the time needed for a half reduction in dehydration/impregnation initial rate; at n = ∞, the process is at equilibrium.

Equation 21 is derived by substituting Eqs. (17)–(19) into Eq. (20):

$$t^i_{1/n} = \frac{k^i_1}{k^i_2}\left(\sqrt{n} - 1\right) \tag{21}$$

So, $t_{1/n}$ represents the ratio between X_{eq} and the initial rate, according to Eqs. (18)–(20).

Using Fick's 2nd law, Crank suggested an equation for diffusion in one dimension in a flat plate, considering an infinite amount of osmotic solution, transient regime and short periods of time, which can be expressed by a simplified equation [7, 18]:

$$\frac{X(t)}{X_{eq}} = 2\sqrt{\frac{D_{eff}(t)^i t}{\pi L^2}} \tag{22}$$

where $D_{eff}(t)^i$ is the instantaneous effective diffusion coefficient. Relating to Eq. (21) with Eqs. (17)–(18) and substitution into Eq. (22), a simple expression is obtained from which $D_{eff}(t)^i$ can be easily calculated at different times.

$$D_{eff}(t)^i = \frac{\pi}{4}\left(\frac{L}{t^i_{1/n}/(\sqrt{n} - 1)^{+1}}\right)^2 t \tag{23}$$

The physical meaning of Peleg's model parameter $t^i_{1/2}$ is the ratio between the constants of Peleg's model. Finally, from Eq. (23) into Eq. (22):

$$\frac{X(t)}{X_{eq}} = \left[\frac{t^i_{1/n}}{t(\sqrt{n} - 1)} + 1\right]^{-1} \tag{24}$$

To compare the diffusivity obtained by the diffusion models with that the one obtained by the model of Barbosa Júnior et al., the average diffusivity was estimated over the time in which X(t) versus $\sqrt{t}$ data were linear, according to Eq. (25) due to the nonuniformity of time intervals.

$$\overline{D_{eff}{}^{i}} = \frac{\int_0^t D_{eff}(t)^i dt}{\int_0^t dt} \tag{25}$$

Estimative standard error (SE) was used to evaluate the best fit to the models:

$$SE = \sqrt{\frac{\sum_{i=1}^{n} (OBS - PRED)^2}{n}} \tag{26}$$

where OBS corresponds to the observed value of water or solid mass and PRED is the predicted value of water or solid mass; n corresponds to the number of observations. The evaluations were conducted by the nonlinear estimation procedure using the software Statistica 5.0® (Statsoft, Tulsa, USA).

3 Results and Discussion

The experimental conditions (see Table 1) carried out to water loss (WL), solids gain (SG), weight reduction (WR) (see Table 2). It also diminished water activity (a_w) and moisture content (U) with respect to the fresh fruit. The differences among the results presented in Table 2 are function of the independent variables (see Table 1). The results of experimental design (CCRD) are showed and discussed as follow.

3.1 Effect of Osmotic Process Variables on the Parameters of Mass Transfer

Table 3 presents the statistical data obtained for the statistical model of each dependent variable, WL, SG, WR, a_w and U. The analysis of variance (ANOVA) with a significance level of 5 %, tested the adequacy of the models using the F test for planning studied and neglecting the non-significant factors [42]. The values of F and the coefficients of correlation ensures the adequacy of the models (Eqs. 27–31, with coded variables X_1, X_2, X_3 and X_4 correspondent to temperature, pressure of vacuum pulse, sodium chloride concentration and sucrose concentration, respectively).

$$\begin{aligned} WL = {} & 33.02 - 1.96X_1 - 0.18X_2 + 2.56X_3 + 4.58X_4 - 0.02X_1{}^2 - 0.07X_2{}^2 - 0.48X_3{}^2 \\ & - 0.13X_4{}^2 - 0.46X_1X_2 - 0.11X_1X_3 + 0.94X_1X_4 + 0.44X_2X_3 - 0.16X_2X_4 - 0.63X_3X_4 \end{aligned} \tag{27}$$

Table 2 Results of experimental design for water loss (WL, %), solid gain (SG, %), weight reduction (WR, %), water activity (a_w) and moisture content (U, kg H_2O/100 kg sample)

Run	WL	SG	WR	a_w	U
1	23.1 ± 0.8	1.7 ± 0.1	18.1 ± 0.9	0.978 ± 0.001	85.99 ± 0.33
2	24.4 ± 1.1	1.5 ± 0.1	19.4 ± 1.2	0.979 ± 0.000	85.80 ± 2.42
3	23.1 ± 3.2	1.5 ± 0.1	17.1 ± 2.4	0.981 ± 0.001	86.33 ± 0.15
4	24.2 ± 1.7	1.5 ± 0.1	18.9 ± 1.9	0.976 ± 0.001	85.50 ± 0.32
5	26.9 ± 1.0	1.9 ± 0.1	19.9 ± 1.2	0.960 ± 0.001	83.25 ± 0.17
6	33.0 ± 3.1	2.7 ± 0.1	24.8 ± 3.8	0.951 ± 0.001	80.22 ± 0.11
7	31.8 ± 2.6	2.4 ± 0.3	24.6 ± 3.1	0.954 ± 0.002	81.80 ± 0.35
8	29.5 ± 1.5	2.3 ± 0.1	24.5 ± 1.2	0.953 ± 0.001	81.51 ± 0.24
9	33.7 ± 2.9	2.5 ± 0.1	27.5 ± 3.6	0.971 ± 0.003	82.54 ± 0.54
10	38.9 ± 0.9	3.1 ± 0.1	31.9 ± 1.2	0.970 ± 0.001	80.04 ± 0.17
11	31.1 ± 5.6	2.1 ± 0.3	27.1 ± 5.2	0.975 ± 0.001	83.20 ± 0.74
12	38.1 ± 1.1	3.1 ± 0.2	32.0 ± 2.8	0.972 ± 0.001	80.09 ± 0.19
13	36.6 ± 3.1	3.2 ± 0.2	28.9 ± 3.9	0.953 ± 0.001	78.66 ± 0.59
14	41.4 ± 2.8	4.2 ± 0.5	34.8 ± 3.5	0.947 ± 0.001	75.78 ± 0.19
15	37.9 ± 2.3	3.0 ± 0.2	30.6 ± 2.8	0.961 ± 0.002	79.82 ± 0.08
16	42.2 ± 1.2	3.1 ± 0.2	35.8 ± 1.5	0.948 ± 0.002	76.81 ± 0.60
17	28.8 ± 1.0	2.3 ± 0.1	19.2 ± 1.2	0.981 ± 0.002	85.76 ± 0.50
18	38.7 ± 0.5	3.4 ± 0.2	31.6 ± 0.6	0.962 ± 0.001	80.96 ± 0.29
19	34.6 ± 1.9	2.6 ± 0.3	26.5 ± 2.3	0.971 ± 0.001	82.96 ± 0.50
20	32.5 ± 1.1	2.5 ± 0.2	23.8 ± 1.4	0.968 ± 0.001	82.25 ± 0.83
21	27.2 ± 1.4	2.7 ± 0.1	17.2 ± 1.8	0.990 ± 0.001	79.82 ± 0.08
22	36.6 ± 1.9	3.3 ± 0.2	28.9 ± 2.4	0.948 ± 0.002	78.25 ± 0.48
23	26.8 ± 1.9	2.6 ± 0.1	16.6 ± 2.4	0.970 ± 0.001	83.78 ± 0.57
24	39.9 ± 2.6	3.4 ± 0.2	33.0 ± 3.2	0.966 ± 0.005	78.83 ± 2.10
25	33.4 ± 1.5	3.0 ± 0.1	24.9 ± 1.9	0.968 ± 0.002	81.75 ± 0.33
26	32.8 ± 0.2	3.0 ± 0.1	24.2 ± 0.2	0.965 ± 0.001	80.55 ± 0.34
27	33.2 ± 1.6	3.1 ± 0.2	24.6 ± 2.0	0.965 ± 0.001	81.11 ± 0.35
28	33.6 ± 1.8	3.1 ± 0.0	24.5 ± 2.1	0.965 ± 0.002	81.20 ± 0.34

$$\begin{aligned} SG = {} & 3.06 + 0.22X_1 - 0.10X_2 + 0.30X_3 + 0.43X_4 - 0.10X_1^2 - 0.20X_2^2 - 0.08X_3^2 \\ & - 0.08X_4^2 - 0.08X_1X_2 + 0.03X_1X_3 + 0.13X_1X_4 - 0.03X_2X_3 + -0.10X_2X_4 - 0.02X_3X_4 \end{aligned} \tag{28}$$

$$\begin{aligned} WR = {} & 24.55 - 2.21X_1 - 0.01X_2 + 2.31X_3 + 4.75X_4 + 0.44X_1^2 + 0.37X_2^2 - 0.13X_3^2 \\ & - 0.30X_4^2 - 0.28X_1X_2 + 0.22X_1X_3 + 0.78X_1X_4 + 0.54X_2X_3 - 0.03X_2X_4 - 0.54X_3X_4 \end{aligned} \tag{29}$$

$$\begin{aligned} a_w = {} & 0.96 - 0.003X_1 - 0.01X_3 - 0.001X_4 - 0.001X_1X_3 + 0.001X_2X_4 \\ & + 0.001X_2X_4 + 0.001X_3X_4 \end{aligned} \tag{30}$$

Table 3 Statistical data for water loss (WL), solid gain (SG), weight reduction (WR), water activity (a_w) and moisture content (U)

	SS	df	AS	Fc	Ft
WL					
Regression	788.7832	9	87.6426	31.95	2.00
Waste	49.3685	18	2.7427		
Lack of fit	49.0310	15			
Pure error	0.3375	3			
Total	838.1517	27			$R^2 = 0.9411$
SG					
Regression	8.4888	6	1.4748	9.6747	3.81
Waste	3.2012	21			
Lack of fit	3.1786	18			
Pure error	0.0226	3	0.0075		
Total	12.0500	27			$R^2 = 0.7918$
WR					
Regression	820.95242	12	68.4127	25.3252	2.02
Waste	40.520384	15	2.7013589		
Lack of fit	40.275284	12			
Pure error	0.2450994	3			
Total	861.4728	27			$r^2 = 0.9530$
a_w					
Regression	0.0031069	7	0.0004	36.5995	2.51
Waste	0.0002428	20	1.156×10^{-5}		
Lack of fit	0.0002373	17			
Pure error	5.5×10^{-6}	3			
Total	0.0033497	27			$r^2 = 0.9276$
U					
Regression	181.29485	7	25.8992	22.5509	2.51
Waste	22.969558	20	1.1484779		
Lack of fit	22.244965	17			
Pure error	0.724593	3			
Total	204.2644	27			$r^2 = 0.8875$

$$U = 81.15 - 1.06X_1 + 0.05X_2 - 1.45X_3 - 1.80X_4 + 0.56X_1^2 + 0.38X_2^2 - 0.50X_3^2 + 0.06X_4^2 - 0.08X_1X_2 - 0.16X_1X_3 - 0.44X_1X_4 + 0.07X_2X_3 + 0.18X_2X_4 + 0.12X_3X_4 \quad (31)$$

The analysis of the coefficients of the Eqs. (27)–(31) shows that:

- The concentration of sucrose (X_4) was the most influent variable on WL, SG, WR and U and the third more influent on a_w. It was directly proportional to WL, SG, WR and inversely proportional to a_w and U. The use of sucrose in ternary solutions sucrose-sodium chloride solution is interesting for tomato dehydration because it could increase WL with lower increasing in SG with respect to a binary solution of sodium chloride [25]. The concentration of sucrose was also the most influent variable for SG, but its coefficient was near the one of the concentration of NaCl (X_3) what makes the concentration of each solute and the concentration of the overall solution more influent than the temperature and vacuum pulse. The direct influence on WR and indirect influence on a_w and U could be seen as consequences of its influence on WL and SG.

- The concentration of sodium chloride (X_3) was the most influent variables a_w and the second one for the other dependent variables, WL, SG, WR and U. Its low molecular weight and high dissociation in ions are the reasons of its low a_w. Moreover, it could cause some structural changes in the cell membrane and presents high permeability in biological materials, diminishing the a_w of the product [53]. The direct influences on WL, SG, WR and inversely proportional influence on U are other consequences [34].
- The temperature (X_1) was the third influent variable for WL, SG, WR and U and the second one for a_w. The greater mobility of the transferred material, the higher permeability and the lower viscosity due to the temperature increase could be pointed as the reasons for the direct influence on SG and inverse influence on U [53]. Consequently, a direct influence on WL and WR was expected in the present work. It was not observed and should be more explored in further works.
- The vacuum pulse (X_2) was the variable with smallest influence on WL, SG, WR and U and with no influence on a_w. The entrance of the solution due to the lower pressure (X_2) is expected to increase the WL, SG and WR with consequent decrease on U [45]. However, the behavior was the opposite for WL and WR. The phenomenon is highly dependent on the structure of the material. The low porosity of the tomatoes could make the action of the vacuum pulse hard [41, 57]. By analyzing the values of the coefficient of X_2 with respect to the other ones, one can see that it was much smaller than the others for WL, WR and U and in the same order for SG. It could be pointed as relevance for SG and low relevance for WL, WR and U.

3.2 Obtaining the Best Process Condition

In a PVOD process, it is essential to achieve high WL. Besides, it is generally desirable to have a final product with low incorporation of the solutes [35].

The optimization condition was performed for the greater values of WL and WR and lower SG, a_w and U. The optimal condition was 40 °C, 56.25 mbar, 7.5 % sodium chloride and 32.5 % sucrose. That condition resulted in 42.2 % of WL, 4.0 % of SG, 36.1 % of WR, a_w of 0.948 and U of 76.5 kg water/100 kg sample.

3.3 Kinetic of Mass Transfer

The kinetics of WL and SG for PVOD of sliced tomatoes in the optimized condition (40 °C, 56.25 mbar, 7.5 % sodium chloride and 32.5 % sucrose) are shown by Fig. 2. The kinetics of OD in the same conditions, except within atmospheric pressure are presented in the same figure. It is possible to see that WL and SG increased with time, mainly in the first 120 min. Moreover, it was experimentally observed a trend to increase WL and SG with vacuum pulse application.

The experimental data were fitted by three models of the literature with coefficient of determination, r^2, and standard error, SE, shown in Table 4. The model of Barbosa Júnior et al. was the one with the best fitness (see Table 4, Figs. 2 and 3) with higher r^2, lower SE and higher proximity between predicted and observed values for both OD and PVOD. The model of Crank was better for OD than the model of Fito and Chiralt and these latter models were similar for PVOD data. The model of Crank and the model of Fito and Chiralt calculate the effective diffusion coefficient as a mean, for the entire duration of the osmotic process. Despite the model of Fito and Chiralt considers the diffusion and hydrodynamic mechanisms during PVOD, the model of Barbosa Júnior et al. achieved better fit, even for PVOD process. The good fit of the model of Barbosa Júnior et al. could be attributed to the more realistic consideration of D_{eff} variation during the osmotic process.

The values of effective diffusion coefficients for all models were in the same order of magnitude and were analogous to those published by several authors. Mercali et al. [35] in an osmotic dehydration of banana with ternary solution, the effective diffusion coefficient of water ranged from 5.19 to 6.47 × 10^{-10} m^2 s^{-1}, the sucrose effective diffusion coefficient between 4.27 and 6.01 × 10^{-10} m^2 s^{-1} and that of NaCl between 4.32 and 5.42 × 10^{-10} m^2 s^{-1}. Silva et al. [46] reported that

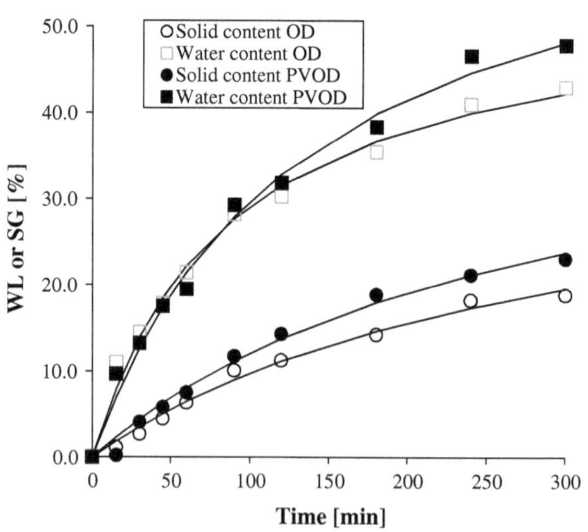

Fig. 2 Experimental and simulated (model of Barbosa Júnior et al.) of kinetics of WL and SG during osmotic dehydration of sliced tomato in ternary solution with (PVOD) and without vacuum pulse (OD)

Table 4 Effective diffusion coefficient (D_{eff}) calculated with model of Crank, model of Fito and Chiralt and model of Barbosa Júnior et al. for osmotic dehydration (OD) and pulsed vacuum osmotic dehydration (PVOD)

Treatments		PVOD	OD
Model of Crank			
WL	D_{eff} ($\times 10^{+10}$ m² s^{-1})	3.77	4.12
	r²	0.9773	0.9832
	SE	0.07	0.06
SG	D_{eff} ($\times 10^{+10}$ m² s^{-1})	1.43	1.64
	r²	0.9598	0.9863
	SE	0.07	0.04
Model of Fito and Chiralt			
WL and SG	D_{eff} ($\times 10^{+10}$ m² s^{-1})	3.98	4.40
	r²	0.9717	0.9790
	SE	0.05	0.04
Model of Barbosa Júnior et al.			
WL	D_{eff} ($\times 10^{+10}$ m² s^{-1})	2.68	3.51
	r²	0.9912	0.9909
	SE	0.02	0.01
SG	D_{eff} ($\times 10^{+10}$ m² s^{-1})	1.52	1.43
	r²	0.9891	0.9902
	SE	0.01	0.01

the effective diffusion coefficient varied from 3.73×10^{-10} to 6.16×10^{-10} m^2 s^{-1} for WL and 1.89×10^{-10} to 5.95×10^{-10} m^2 s^{-1} for SG during the osmotic dehydration of pineapples in sucrose binary solution and ternary sucrose-calcium lactate solution. Mundada et al. [36] reported that the effective diffusion coefficient ranged from 2.72×10^{-10} to 5.12×10^{-10} m^2 s^{-1} and from 1.47×10^{-10} to 5.15×10^{-10} m^2 s^{-1} for WL and SG, respectively, during osmotic dehydration of pomegranate arils in sucrose solutions. Ito et al. [29] reported ranges from 5.39×10^{-10} to 9.75×10^{-10} m^2 s^{-1} for water loss and 2.74×10^{-10} to 1.40×10^{-9} m^2 s^{-1} for solid gain during pulsed vacuum osmotic dehydration of mango slices in sucrose solutions.

With the exception of the model of Fito and Chiralt, where the D_{eff} value for water and solid are the same, all tested models followed the same trends: values of D_{eff} for WL were higher than D_{eff} values for SG. It is in accordance to the experimental data (see Fig. 2) and is due to the ease of diffusion of water molecules. Although the fitness was good for all the models, the values of D_{eff} for OD was higher than for PVOD, what is not in accordance to the experimental data, except for SG in a PVOD fitted by the model of Barbosa Junior et al. It could be attributed to the small difference of the kinetics between OD and PVOD data and the error of fitness of the models.

Table 5 presents parameters of the model of Barbosa Júnior et al. for WL and SG in OD and PVOD experiments. By analysing the data of the predicted values of WL and SG in the equilibrium, WL_∞ and SG_∞ and the half time of dehydration/impregnation rate ($t_{1/2}$) it is possible to note that the time of 300 min, choose for this

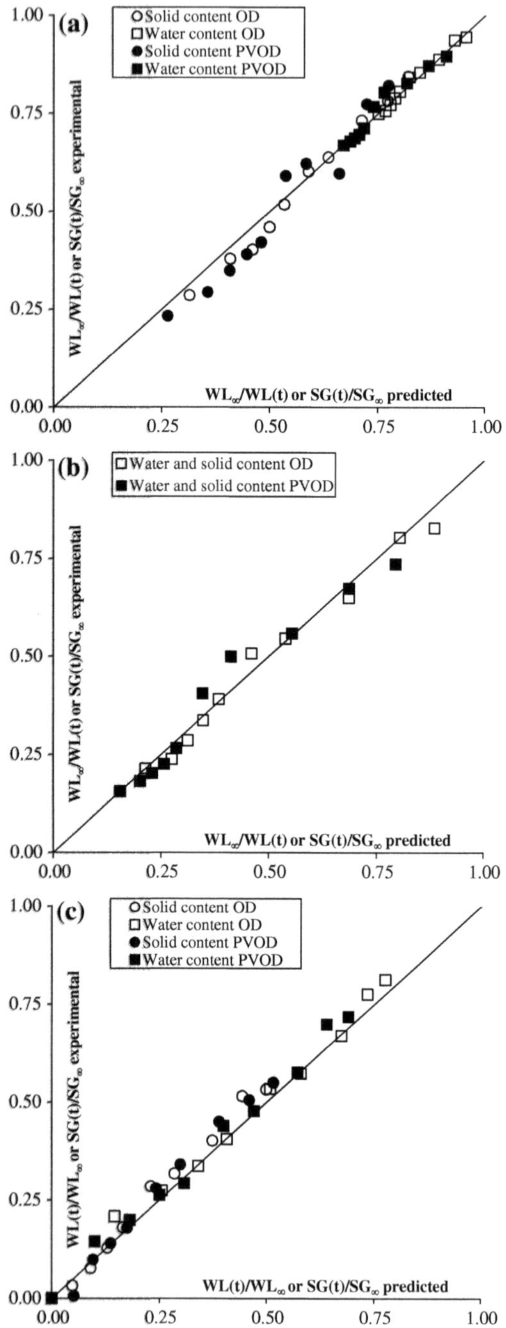

Fig. 3 Fitting capability of the model of Crank (**a**), model of Fito and Chiralt (**b**) and model of Barbosa Júnior et al. (**c**) for WL and SG kinetics during osmotic dehydration with and without pulsed vacuum of sliced tomato in ternary solution

Table 5 Mass transfer kinetics parameters for water loss (WL) and solid gain (SG) obtained using the model of Barbosa Júnior et al.

Treatment	WL (kg water/100 kg sample)				SG (kg solid/100 kg sample)			
	${}_{a}WL_{\infty}$	${}_{c}1/k_1'$	${}_{d}1/k_2'$	${}_{e}t_{1/2}'$ (min)	${}_{b}SG_{\infty}$	${}_{c}1/k_1''$	${}_{d}1/k_2''$	${}_{e}t_{1/2}''$ (min)
OD	52.87	0.1044	54.47	36.02	35.44	0.0217	39.14	124.50
PVOD	66.64	0.0862	69.76	55.83	41.98	0.0273	46.08	116.56

${}_{a}WL_{\infty}$: experimental value of WL at equilibrium (kg water/100 kg sample), ${}_{b}SG_{\infty}$: experimental value of solid gain at equilibrium (kg solid/100 kg sample). ${}_{c}1/k_1'$ calculated initial rate (mg water/1 g sample/minute), .${}_{c}1/k_1''$: calculated initial rate (mg solid/g sample/minute). ${}_{d}1/k_2'$ WL at equilibrium calculated using the model (kg water/100 kg sample), $1/k_2''$: SG at equilibrium calculated using the model (kg solid/100 kg sample). ${}_{e}t_{1/2}'$ 'half-life' of dehydration and $t_{1/2}''$: 'half-life' of impregnation rate

work was far from the equilibrium condition for WL and especially for SG. It is interesting to note that although most of the values of predicted D_{eff} was not in accordance with the experimental data (see Fig. 2), the predicted values of WL_{∞} and SG_{∞} were higher for PVOD, in accordance with the experimental kinetics.

4 Conclusion

The concentration of sucrose was the most influent variable on the mass transfer parameters WL, SG and WR of the osmotic dehydrated of tomato slices in ternary solution of sucrose-NaCl.

Although vacuum pulse presented influence on WL, WR and moisture reduction, its influence was more evident for SG.

The optimum conditions for maximum water loss and weight reduction, minimal water activity and moisture content was successfully obtained by the response surface methodology. The result was 42.22 % of WL, 4 % of SG with WR equal to 36.06 %, a_w of 0.948 and moisture content of 76.5 kg water/100 kg sample.

A model that considers variable diffusivity carries out to better fitness of osmotic dehydration with or without vacuum pulse.

Acknowledgments The authors would like to thank the economical support CNPq (National Council for Scientific and Technological Development), FAPEMIG (State of Minas Gerais Research Foundation) and CAPES (Coordination for the Improvement of Higher Education Personnel).

References

1. Abbasi Souraki, B., Ghavami, M., Tondro, H.: Comparison between continuous and discontinuous method of kinetic evaluation for osmotic dehydration of cherry tomato. J. Food Process. Preserv. **38**(6), 2167–2175 (2014)
2. Al-Harahsheh, M., Al-Muhtaseb, A.H., Magee, T.R.: Microwave drying kinetics of tomato pomace: effect of osmotic dehydration. Chem. Eng. Process. Process Intensif. **48**(1), 524–531 (2009)

3. An, K., Li, H., Zhao, D., et al.: Effect of osmotic dehydration with pulsed vacuum on hot-air drying kinetics and quality attributes of cherry tomatoes. Dry. Technol. **31**(6), 698–706 (2013)
4. AOAC-Official Methods of Analysis: Association of Official Analytical Chemists, Washington (2007)
5. Arballo, J.R., Bambicha, R.R., Campañone, L.A., et al.: Mass transfer kinetics and regressional-desirability optimisation during osmotic dehydration of pumpkin, kiwi and pear. Int. J. Food Sci. Technol. **47**(2), 306–314 (2012)
6. Azarpazhooh, E., Ramaswamy, H.S.: Evaluation of diffusion and Azuara models for mass transfer kinetics during microwave-osmotic dehydration of apples under continuous flow medium-spray conditions. Dry. Technol. **28**(1), 57–67 (2010)
7. Azuara, E., Beristain, C.I., Garcia, H.S.: Development of a mathematical model to predict kinetics of osmotic dehydration. J. Food Sci. Technol. **29**, 239–242 (1992)
8. Barbosa Júnior, J.L., Cordeiro, M.M., Hubinger, M.D.: Mass transfer kinetics and mathematical modelling of the osmotic dehydration of orange-fleshed honeydew melon in corn syrup and sucrose solutions. Int. J. Food Sci. Technol. **48**(12), 2463–2473 (2013)
9. Borges, S.V., Mancini, M.C., Corrêa, J.L.G., et al.: Drying kinetics of bananas by natural convection: influence of temperature, shape, blanching and cultivar. Cienc. Agrotec. **35**(8), 368–376 (2011)
10. Borges, S.V., Mancini, M.C., Corrêa, J.L.G., et al.: Secagem de fatias de abóboras (*Curcubit moschata*, L.) por convecção natural e forçada. Cienc. Tecnol. Aliment. **28**, 245–251 (2008)
11. Checmarev, G., Casales, M.R., Yeannes, M.I., et al.: Mass transfer modeling during osmotic dehydration of chub mackerel (*Scomber japonicus*) slices in salt and glycerol solution at different temperatures. J. Food Process. Preserv. **38**(4), 1599–1607 (2014)
12. Chiralt, A., Talens, P.: Physical and chemical changes induced by osmotic dehydration in plant tissues. J. Food Eng. **67**(1–2), 167–177 (2005)
13. Corrêa, J.L.G., Batista, M.B., Raimundo, A., et al.: Analysis of osmotic dehydration variables: influence on tomato (*Licopersicon esculentum*) drying. Bol. CEPPA **25**(2), 315–327 (2007)
14. Corrêa, J.L.G., Dev, S.R.S., Gariepy, Y., et al.: Drying of pineapple by microwave-vacuum with osmotic pretreatment. Dry. Technol. **29**(13), 1556–1561 (2011)
15. Corrêa, J.L.G., Ernesto, D.B., Alves, J.G.L.F., et al.: Optimisation of vacuum pulse osmotic dehydration of blanched pumpkin. Int. J. Food Sci. Technol. **49**(9), 2008–2014 (2014)
16. Corrêa, J.L.G., Pereira, L.M., Vieira, G.S., et al.: Mass transfer kinetics of pulsed vacuum osmotic dehydration of guavas. J. Food Eng. **96**(4), 498–504 (2010)
17. Corzo, O., Bracho, N.: Application of Weibull distribution model to describe the vacuum pulse osmotic dehydration of sardine sheets. LWT—Food Sci. Technol. **41**(6), 1108–1115 (2008)
18. Crank, J.: The Mathematics of Diffusion. Carendon press, Oxford (1975)
19. Czerner, M., Yeannes, M.I.: Brining kinetics of different cuts of anchovy (*Engraulis anchoita*). Int. J. Food Sci. Technol. **45**(10), 2001–2007 (2010)
20. Deng, Y., Zhao, Y.: Effect of pulsed vacuum and ultrasound osmopretreatments on glass transition temperature, texture, microstructure and calcium penetration of dried apples (Fuji). LWT—Food Sci. Technol. **41**(9), 1575–1585 (2008)
21. Fante, C., Corrêa, J., Natividade, M., et al.: Drying of plums (*Prunus sp*, cv Gulfblaze) treated with KCl in the field and subjected to pulsed vacuum osmotic dehydration. Int. J. Food Sci. Technol. **46**(5), 1080–1085 (2011)
22. FAOSTAT: FAO Statistics Division. http://faostat.fao.org/site/339/default.aspx (2011). Accessed 12 Dec 2014
23. Ferrari, C.C., Sarantópoulos, C.I.G.L., Carmello-Guerreiro, S.M., et al.: Effect of osmotic dehydration and pectin edible coatings on quality and shelf life of fresh-cut melon. Food Bioprocess. Technol. **6**(1), 80–91 (2013)
24. Fito, P., Chiralt, A.: An approach to the modeling of solid food-liquid operations: application to osmotic dehydration. In: Fito, P., Ortega-Rodriguez, E., Barbosa-Canovas, G. (eds.) Food Engineering 2000, pp. 231–252. Chapman and Hall, New York (1977)

25. Heredia, A., Barrera, C., Andrés, A.: Drying of cherry tomato by a combination of different dehydration techniques. Comparison of kinetics and other related properties. J. Food Eng. **80** (1), 111–118 (2007)
26. Heredia, A., Peinado, I., Barrera, C., et al.: Influence of process variables on colour changes, carotenoids retention and cellular tissue alteration of cherry tomato during osmotic dehydration. J. Food Compos. Anal. **22**(4), 285–294 (2009)
27. Heredia, A., Peinado, I., Rosa, E., et al.: Effect of osmotic pre-treatment and microwave heating on lycopene degradation and isomerization in cherry tomato. Food Chem. **123**(1), 92–98 (2010)
28. Heredia, A., Peinado, I., Rosa, E., et al.: Volatile profile of dehydrated cherry tomato: Influences of osmotic pre-treatment and microwave power. Food Chem. **130**(4), 889–895 (2012)
29. Ito, A.P., Tonon, R.V., Park, K.J., et al.: Influence of process conditions on the mass transfer kinetics of pulsed vacuum osmotically dehydrated mango slices. Dry. Technol. **25**(10), 1769–1777 (2007)
30. Kejing, A., Hui, L., Dandan, Z., et al.: Effect of osmotic dehydration with pulsed vacuum on hot-air drying kinetics and quality attributes of cherry tomatoes. Dry. Technol. **31**(6), 698–706 (2012)
31. Li, H., Zhao, C., Guo, Y., et al.: Mass transfer evaluation of ultrasonic osmotic dehydration of cherry tomatoes in sucrose and salt solutions. Int. J. Food Sci. Technol. **47**(5), 954–960 (2012)
32. Mabellini, A., Ohaco, E., Márquez, C., et al.: Calculation of the effective diffusion coefficients in drying of chemical and mechanical pretreated rosehip fruits (*Rosa eglanteria* L.) with selected sass transfer models. Int. J. Food Eng. **9**(4), 481–486 (2013)
33. Marques, G.R., Borges, S.V., Botrel, D.A., et al.: Spray drying of green corn pulp. Dry. Technol. **32**(7), 861–868 (2014)
34. Mayor, L., Moreira, R., Chenlo, F., et al.: Osmotic dehydration kinetics of pumpkin fruits using ternary solutions of sodium chloride and sucrose. Dry. Technol. **25**(10), 1749–1758 (2007)
35. Mercali, G.D., Marczak, L.D.F., Tessaro, I.C., et al.: Evaluation of water, sucrose and NaCl effective diffusivities during osmotic dehydration of banana (*Musa sapientum*, shum.). LWT—Food Sci. Technol. **4**(4), 82–91 (2011)
36. Mundada, M., Hathan, B.S., Maske, S.: Mass transfer kinetics during osmotic dehydration of pomegranate arils. J. Food Sci. **76**(1), 31–39 (2011)
37. Ochoa-Martınez, C.I., Ramaswamy, H.S., Ayala-Aponte, A.: A comparison of some mathematical models used for the prediction of mass transfer kinetics in osmotic dehydration of fruits. Dry. Technol. **25**(10), 1613–1620 (2007)
38. Ozuna, C., Álvarez-arenas, T.G., Riera, E., et al.: Influence of material structure on air-borne ultrasonic application in drying. Ultrason. Sonochem. **21**(3), 1235–1243 (2014)
39. Palou, E., Lopez-Malo, A., Argaiz, A., et al.: Use of Peleg's equation to osmotic concentration of papaya. Dry. Technol. **12**(4), 965–978 (1994)
40. Porciuncula, B.D., Zotarelli, M.F., Carciofi, B.A., et al.: Determining the effective diffusion coefficient of water in banana (Prata variety) during osmotic dehydration and its use in predictive models. J. Food Eng. **119**(3), 490–496 (2013)
41. Ramallo, L.A., Hubinger, M.D., Mascheroni, R.H.: Effect of pulsed vacuum treatment on mass transfer and mechanical properties during osmotic dehydration of pineapple slices. Int. J. Food Eng. **9**(4), 403–412 (2013)
42. Rodrigues, M.I., Iemma, A.F.: Experimental Design and Process Optimization. Casa do Espírito Amigo Fraternidade, Campinas (2012)
43. Sharma, R., Gupta, A., Abrol, G.S., et al.: Value addition of wild apricot fruits grown in North-West Himalayan regions-a review. J. Food Sci. Technol. **51**(11), 2917–2924 (2014)
44. Shi, J.X., Maguer, L., Wang, L., et al.: Application of osmotic treatment in tomato processing: effect of skin treatments on mass transfer in osmotic dehydration of tomatoes. Food Res. Int. **30**(9), 669–674 (1998)

45. Shi, X.Q., Fito, P., Chiralt, A.: Influence of vacuum treatment on mass transfer during osmotic dehydration of fruits. Food Res. Int. **28**(5), 445–454 (1995)
46. Silva, K.S., Fernandes, M.A., Mauro, M.A.: Effect of calcium on the osmotic dehydration kinetics and quality of pineapple. J. Food Eng. **134**, 37–44 (2014)
47. Silva, M.A.C., Corrêa, J.L.G., Da Silva, Z.E.: Application of inverse methods in the osmotic dehydration of acerola. Int. J. Food Sci. Technol. **45**(12), 2477–2484 (2010)
48. Silva, M.A., Corrêa, J.L.G.: Academic research on drying in Brazil 1970–2003. Dry. Technol. **23**(7), 1345–1359 (2005)
49. Silva, M.A.C., Corrêa, J.L.G., Da Silva, Z.E.: Drying kinetics of west indian cherry: influence of osmotic pretreatment. Bol. CEPPA **29**(2), 193–202 (2011)
50. Silva, M.A.C., Silva, Z.E., Mariani, V.C., et al.: Mass transfer during the osmotic dehydration of West Indian cherry. LWT—Food Sci. Technol. **45**(2), 246–252 (2012)
51. Da Silva, W.P., Amaral, D.S., Duarte, M.E.M., et al.: Description of the osmotic dehydration and convective drying of coconut (*Cocos nucifera* L.) pieces: a three-dimensional approach. J. Food Eng. **115**(1), 121–131 (2013)
52. Telis, V.R.N., Murari, R.C.B.D.L., Yamashita, F.: Diffusion coefficients during osmotic dehydration of tomatoes in ternary solutions. J. Food Eng. **61**(2), 253–259 (2004)
53. Tonon, R.V., Baroni, A.F., Hubinger, M.D.: Osmotic dehydration of tomato in ternary solutions: Influence of process variables on mass transfer kinetics and an evaluation of the retention of carotenoids. J. Food Eng. **82**(4), 509–517 (2007)
54. Udomkun, P., Mahayothee, B., Nagle, M., et al.: Effects of calcium chloride and calcium lactate applications with osmotic pretreatment on physicochemical aspects and consumer acceptances of dried papaya. Int. J. Food Sci. Technol. **49**(4), 1122–1131 (2014)
55. Velickova, E., Winkelhausen, E., Kuzmanova, S.: Physical and sensory properties of ready to eat apple chips produced by osmo-convective drying. J. Food Sci. Technol. **51**(12), 3691–3701 (2014)
56. Viana, A.D., Corrêa, J.L.G., Justus, A.: Optimisation of the pulsed vacuum osmotic dehydration of cladodes of fodder palm. Int. J. Food Sci. Technol. **49**(3), 726–732 (2014)
57. Vieira, G.S., Pereira, L.M., Hubinger, M.D.: Optimisation of osmotic dehydration process of guavas by response surface methodology and desirability function. Int. J. Food Sci. Technol. **47**(1), 132–140 (2012)
58. Vinha, A.F., Alves, R.C., Barreira, S.V.P., et al.: Effect of peel and seed removal on the nutritional value and antioxidant activity of tomato (*Lycopersicon esculentum* L.) fruits. LWT—Food Sci. Technol. **55**(1), 197–202 (2014)
59. Yadav, A.K., Singh, S.V.: Osmotic dehydration of fruits and vegetables: a review. J. Food Sci. Technol. **51**(9), 1654–1673 (2014)
60. Zhao, D., Zhao, C., Tao, H., et al.: The effect of osmosis pretreatment on hot-air drying and microwave drying characteristics of chili (*Capsicum annuum* L.) flesh. Int. J. Food Sci. Technol. **48**(8), 1589–1595 (2013)

MIX
Papier aus verantwortungsvollen Quellen
Paper from responsible sources
FSC® C105338

If you have any concerns about our products,
you can contact us on
ProductSafety@springernature.com

In case Publisher is established outside the EU,
the EU authorized representative is:
Springer Nature Customer Service Center GmbH
Europaplatz 3, 69115 Heidelberg, Germany

Printed by Libri Plureos GmbH
in Hamburg, Germany